COMPUTATIONAL TECHNIQUES IN OPERATIONS RESEARCH

COMPUTER LANGUAGE & PROGRAMMING SERIES

Editor in Chief:

Dr. Brian Rudall

Senior Lecturer, Dept of Computer Science,
University of North Wales, Bangor, UK

Further titles to be announced

COMPUTATIONAL TECHNIQUES IN OPERATIONS RESEARCH

A. M. Andrew Ph.D.

ABACUS PRESS

Tunbridge Wells, Kent

&

Cambridge, Mass

First published in 1985 by

ABACUS PRESS
Abacus House,
Tunbridge Wells,
Kent, TN4 0HU, UK

and

ABACUS PRESS
PBS, PO Box 643,
Cambridge,
Mass 02139, USA

Printed by Antony Rowe Ltd., Chippenham

Second Impression 1986

Library of Congress Cataloguing-in-Publication Data

Andrew, A. M.
Computational techniques in operations research.
(Computer language and programming series)
Bibliography: p.
Includes indexes.
1. Operations research – Data processing.
I. Title. II. Series.
T57.6.A55 1985 011.4'24'0285 85-9110
ISBN 0-85626-425-3

British Library Cataloguing in Publication Data

Andrew, A. M.
Computational techniques in operations research
– (Computer language and programming series)
1. Operations research – Data processing
I. Title II. Series
001.4'24'02854 T.57.6
ISBN 0-85626-425-3

CONTENTS

SERIES PREFACE

This series has two aims.

First to provide a detailed background knowledge of computers to all who use computing machines. It is a theme of each of these monographs that we need to have a good understanding of computers and the way in which they function, if we are to fully appreciate their increasingly important role in society.

Secondly the series aims to consider the material impact of computers by discussing some of their principal applications, particularly those concerned with practical problems.

Several of the books in this series will concentrate on providing a high-level introduction to computing machines and their software with particular attention paid to computer languages and to the art of programming. Others will be concerned with the use of computer programming languages for specific applications. Later volumes will consider the well-defined topics that are associated with current applications and with the development of the 'man-machine interface' which provides the tools for problem-solving on the new generation of computers.

There has been an 'information explosion' in the field of computing, especially in the design, development and implementation of programming languages. Much of this has occurred because of the recent developments in microelectronics which have profoundly affected our approach to computers. It is essential to realise that computing machines have been developed throughout the ages and have reflected, in their design, manufacture and application, the technology of their period. Consequently the most recent developments in computer languages and programming and their use in specific application areas will be highlighted.

The authors of the books chosen for inclusion in this series, are all well-known experts in their particular fields. Each monograph should become standard reading and study material for all those who use computers.

B. H. RUDALL
University College of North Wales, UK

PREFACE

The material presented here is adapted from lectures I gave as part of the degree course in Cybernetics in the University of Reading. I am indebted to Professor P. B. Fellgett for helpful discussions about course content, and before that to Professor Stafford Beer and other colleagues in Sigma (Science in General Management) Ltd., for introducing me to practical Operations Research.

In preparing the book I have received much help from my wife Joyce. No doubt it still has errors, ambiguities and inelegant wordings, but they are much fewer than they were in the preliminary draft which she checked.

Rather to my surprise, Joyce's perusal of the book produced some hilarity. Several of the techniques involve rather preposterous stratagems by which problem situations are forced to be mathematically tractable. These include the introduction of artificial variables which are then eliminated, dummy destinations supplied without movement of goods, dummy football fixtures which are not played, and dummy activities completed in zero time.

I have always felt that science can and should be fun, so I am happy if the book is read for its entertainment value. I hope it also has utility in introducing readers to these valuable techniques and also, perhaps, giving some instruction, mainly by example, in computer programming. It also provides a number of programs which may be used as they stand or may form the basis of versions tailor-made to readers' requirements. I think I have made the topics accessible to a wide range of readers, from the interested layman to mathematicians, and specialists in computing, who have not previously become acquainted with O.R. methods.

A.M. ANDREW
Viable Systems, Chillaton, Devon

Chapter 1

INTRODUCTION TO O.R. AND LANGUAGES

OPERATIONS RESEARCH

The term Operations Research came into use during the Second World War to denote the application of quantitative methods in management. Under wartime conditions it was important to make the best possible use of limited resources, and the need for efficient management was particularly obvious. It is also obvious that, under any conditions, there is no merit in waste and any actions taken should be planned to make the best possible use of resources. Of course, what is meant by 'best' is not always obvious, and account has to be taken of the cost of the computation needed to decide how to act. Rough-and-ready solutions may be appropriate when the possible costs or payoffs are not great.

The techniques which will be described are all concerned with optimisation. That is to say, they allow the planning of some activity so as either to maximise something which may be termed 'profit' or to minimise something which may be termed 'cost'. Although there is a large literature on Operations Research, the techniques are not as widely known as they deserve to be, even among mathematicians and experts on other aspects of computation. This is somewhat surprising since the techniques have applications in everyday living and in computational tasks having little or no connection with management. For example, an application of Linear Programming to networks of threshold elements or 'model neurons' is mentioned in Chapter 3.

It will be seen that the techniques are not particularly difficult to understand and apply. Most of them can fairly readily be applied without the use of a computer, in simple situations. However, they soon become tedious when more complicated situations are treated and the emphasis here will be on the use of electronic rather than manual computation. The computer programs presented are fairly short and can be accommodated on small machines.

The presentation of computer programs should be valuable even to a reader who does not have access to computing machinery. They are in one or other of the high-level languages BASIC and PASCAL, and apart from their potential usefulness when entered on machines they provide a clear and rigorous presentation of the methods for human perusal.

LIMITATIONS OF NUMERICAL OPTIMISATION

Despite what was said above it is important not to over-sell the usefulness of these computational techniques and the somewhat blinkered view of the scope of O.R. which they may foster. Their use often depends on a rather severe idealisation of the real-world situation. One of the difficulties is that the data on which the optimisation has to be based are frequently uncertain, and it is often impossible even to guess with much confidence at probability distributions for numerical values. The uncertainty need not be restricted to numerical data; the qualitative structure of an existing or future situation is often uncertain.

What may be a more serious difficulty than the uncertainty of the data is the need to represent all problems in terms of maximising numerical criteria. (Minimising a criterion is equivalent to maximising its negative, so need not be separately mentioned). People often think of business as a matter of maximisation of monetary profits, and business men are described as hard-headed with the implication that profit is their only goal. In fact, even the most hard-headed of businessmen pay attention to a great many non-quantifiable potential outcomes of their decisions - they would not be very successful in their businesses if they did not.

Decisions in business, as in everyday living, are made with regard to such non-quantifiable considerations as the maintenance of good relations with the public who are the potential customers, with the work force and with other businesses and official bodies. Also, the owner or manager of a business may prefer, for his own personal satisfaction, to behave with social rsponsibility and for example to make a sound product even where the market conditions are such that he could palm off an inferior one on his customers.

It is in fact possible to attribute many ills of modern society to the adoption of attitudes encouraged by a simplistic interpretation of Operations Research, with the consequent dehumanising of relationships. This is a big topic which will not be further developed here. It has to be acknowledged that decisions have to be made in the context of a complex interactive environment (see, for example, Beer [1]) and that numerical optimisation techniques are only a part of the armoury of a competent O.R. specialist. They are not a substitute for managerial judgement. On the other hand, there are some important contexts in which the numerical techniques are directly and unreservedly applicable. In others, they have value as an aid to human decision-making - a manager with years of experience is a better one if he takes account of the results of numerical optimisation of simplified situations than if he operates entirely informally. Of course, these considerations apply not only to large-scale management with connotations of Corridors of Power, but also to 'management' in a wide sense including much of everyday living.

THE PROGRAMMING LANGUAGE BASIC

BASIC is an example of a high-level language. It is intended to be particularly easy to learn and to use, and it is the language most widely used for programming small machines. The name is an acronym, being the initials of 'Beginners All-purpose Symbolic Instruction Code'. It is so well known that the brief introduction given here will be largely superfluous for most readers. However, it is probably worth including for the sake of completeness and to indicate the features of the precise version, or 'dialect' of BASIC which is used. Perhaps as a consequence of its early introduction and widespread use, the BASIC language is not well standardised and there can be quite significant differences between the various dialects. The programs in BASIC presented here are in the version of the language which runs on the Sinclair ZX SPECTRUM.

HIGH-LEVEL LANGUAGES

In the very early days of electronic digital computers, programming had to be done in machine language. That is to say, the user had to prepare the programs in essentially the form in which they would ultimately be stored in the computer memory to determine the sequence of elementary operations needed in the calculation. The writing of programs directly in machine language is extremely tedious and prone to error.

Some of the tedium results from the need for the programmer to keep track of which storage location in the computer memory he is using for each purpose, and of the numerical codes corresponding to the different types of elementary operation of the computer. These are tasks of the clerical kind which we hope to eliminate by employing computers, so an obvious improvement is to let the programmer write the programs in a way which is easier for him because he can refer to storage locations by groups of characters having some meaning for him, such as 'total' or 'count'. To refer to locations occupied by program instructions he is likely to use words like 'start' or 'repeat'. He can also be allowed to refer to computer operations by groups of characters like ADD, SUB or JMP (short for 'jump'). A fairly simple computer program termed an assembler can convert the program written in this way to an actual machine-language program, and such program-writing is called assembly-language programming.

More recent developments make it possible to write programs in ways which are even better suited to the human and further removed from the machine's internal language. The languages which are then used are termed high-level languages and they allow programming to be done by people who have no detailed knowledge of the internal working, and hence the machine language, of the computer. The program which translates the high-level language program into a machine-language one is termed a compiler. It is possible for the same program in a high-level language to be compiled and run on a number of quite different machines - the programming is machine-independent.

It is not necessary to compile a program in its entirety before starting to run it; it can be stored in its high-level, or source, form and translated a little bit at a time as needed to let it run. A system which allows this kind of operation is termed an interpretive system. Programs in BASIC may be run either interpretively or using a compiler. They run much more quickly if they have been compiled, but program development can usually be done more conveniently using an interpretive system. An interpretive system is clearly rather inefficient when, as in most programs, many of the instructions are obeyed many times over during one run of the program, since each re-use of an instruction requires its re-interpretation. However, for BASIC programming the cost of this in run time is often thought to be worth while in view of the ease with which programs can be modified without the need for total re-compilation.

For the present purpose it is immaterial whether the BASIC programs are compiled or run interpretively - the form of the program is the same for either.

DETAILS OF BASIC

In programs in high-level languages, as in assembly-language programs, letters of the alphabet such as x or y, or groups of symbols starting with a letter, such as 'total' or 'x2', are used in a way which looks very similar to how letter symbols are used in algebra. In the programming use, however, the groups of symbols (sometimes called names or identifiers) refer to areas of the computer's memory. These areas, or locations, most commonly hold numbers, though they may hold other types of item such as text strings. The usual convention in BASIC is that an identifier denotes a location capable of holding a text string (or an array of text strings) if it ends with a dollar sign, as 'a$', but otherwise it denotes a location able to hold a number (or an array of numbers, but for the time being arrays are ignored).

Each BASIC statement or instruction is introduced

by a special word which indicates the type of statement it is. The most common of these special words is LET, and a statement of the form:

```
LET x=5
```

has the effect of setting the numerical value in the location denoted by x to the value 5. Similarly, the value stored in the location denoted by y is copied to that denoted by x in response to the statement:

```
LET x=y
```

and arithmetic operations are carried out if an expression, rather than a simple variable, appears to the right of the equals sign, as in:

```
LET x=y+z
LET x=x+1
```

In the second of these, the name x appears on both sides of the equals sign. This is quite in order, and the appearance (or appearances) on the right-hand side refer to the value held as x before the instruction is obeyed. The effect of the statement as shown is to increment the stored value by unity.

Each line of a BASIC program starts with a line number. The usual arrangement is that, if a statement is typed-in without a line number it is obeyed as soon as the carriage-return is pressed to indicate completion of the line. This means that a computer with a BASIC system running on it can be used as a rather fancy desk calculator. To obtain the result of multiplying 123 by 456 one can simply type in:

```
PRINT 123*456
```

and when the carriage return (or 'enter') key is pressed the computer does the calculation and displays the answer of 56088. The asterisk is used as a multiplication sign, and the slash (/) for division, since the more usual multiplication and division signs are not available on most keyboards. However, if the statement is typed in with a line number, as:

```
10 PRINT 123*456
```

nothing is seen to happen immediately. The statement

is stored to become part of a program, and will only be obeyed when the whole program is activated by giving the command RUN. All the lines in the program must be numbered differently, and before being obeyed they are ordered according to the numbering. This is not necessarily the order in which they were entered. It is usual to number the lines 10, 20, 30 ... rather than 1, 2, 3 ..., so that intermediate statements numbered, say, 15 or 25, can be introduced if this is found to be necessary because something was forgotten in the first place. The line numbers can be referred to in other statements in the program, as in:

```
30 GOTO 10
40 IF x=1 THEN GOTO 10
```

In some versions of BASIC there can be only one statement per line. Other versions allow a number of statements to follow one another, separated by colons, and they all count as one program line, with one line number. They form one line in the sense that no carriage-return, or 'enter', has been inserted through the keyboard, but when the program is displayed on the screen, or printed, the program line may occupy several lines of text. The version of BASIC provided on the Sinclair ZX SPECTRUM allows a line to hold a number of statements. Of course, a GOTO statement can refer only to the start of a line since it uses a line number. (SPECTRUM BASIC does actually allow jumps to statements which are not at the beginnings of lines, but it is done using a rather clumsy combination of instructions. In practice it is always preferable to divide the sequence of statements so that the target one is at the beginning of a line.)

The facility whereby several statements can appear in one line, separated by colons, is particularly useful in IF or conditional statements. These are of the general form:

```
10 IF a>0 THEN statement1: statement2: ...
```

where 'a>0' is one example of a condition to be tested. If the condition yields 'true', the statements following THEN are all obeyed, but if it yields 'false' they are all skipped. The fact that the switching in-and-out of a section of program is determined by one test of a condition gives the language something of the 'structured programming' quality associated with the ALGOL-type languages, including PASCAL. There are in fact dialects of BASIC which embody much more structure than this and might be said to be ALGOL-type.

Other features of BASIC, such as the means of producing program loops and of using arrays, will be introduced later, when they can be illustrated in non-trivial programs. A small BASIC program will be shown at this stage to consolidate what has been introduced so far. It is necessary, first, to introduce three further types of statement, all of them fairly self-explanatory. A statement beginning with the word INPUT, such as:

```
10 INPUT x
```

has the effect of showing a 'prompt', usually a question mark, on the computer screen, and halting operation until the user types in a numerical value (followed by 'enter') to be assigned to the storage location denoted by x. If it had been:

```
10 INPUT a$
```

the input would have been treated as a text string rather than as a number. One statement of this kind can require several items of data to be inserted, as in:

```
10 INPUT a$,x,b$,y
```

A statement beginning with REM has no effect on the operation of the program - it simply provides a place where the programmer can insert explanatory remarks. The statement STOP does just what it says and terminates the running of a program. The function INT is a means of obtaining the integer part of a number - for instance, if x holds the value 12.34, INT(x) has the value 12.

The following is a program to accept a number as input and to print out "Non-integer" if it is not an integer, or to print "Odd" or "Even" as appropriate if it is. It should be mentioned that the combination of symbols <> is used to mean 'not equal to', and that, in a PRINT statement, a text string enclosed in quotes is printed as it stands (without the quotes). It will be seen that the program listing takes new lines of display at rather odd places within a program line - this is in accordance with the display of the Sinclair SPECTRUM:

```
10 REM Parity program
20 INPUT x
30 IF x <> INT (x) THEN PRINT "N
on-integer": STOP
40 IF x = 2*INT (x/2) THEN PRINT
"Even": STOP
50 PRINT "Odd": STOP
```

If, after inserting the above into a computer, it was decided to have more information printed out, the following could be typed in:

```
22 IF x = 0 THEN PRINT "Zero": STOP
24 IF x > 0 THEN PRINT "Positive": GOTO 30
26 PRINT "Negative"
```

The BASIC system sorts all the program lines into their numerical order, so the new ones are placed between lines 20 and 30 of the program as first inserted. In the expanded program, the line to be obeyed after line 24 may be either line 30 if the condition x > 0 yields 'true', or the next line in sequence, namely 26, if it yields 'false'.

THE PROGRAMMING LANGUAGE PASCAL

The BASIC language is similar to its widely-used precursor FORTRAN in relying on the labelling of statements (the line numbers being effectively labels) and then the use of GOTO statements to produce transfers of control during program execution. If there are more than a very few GOTO statements in a program it can be almost impossible to trace out all the possible paths through it by simple inspection of a print-out, or listing, of it. It may be necessary to draw a flow diagram in which nodes correspond to program statements and are marked with the same labels. They are joined by lines marked with arrow-heads to show how control may be transferred during execution. Introductory texts on programming often give the clear direction:

"Start by drawing a flow diagram".

If the programming is to be in FORTRAN or BASIC this is usually good advice.

There is, however, another approach to programming. The intention of the ALGOL-type languages is that the structure of the computing task should correspond to that of the program, in a way which is self-evident without the need for a flow diagram as an intermediary. One of the tricks used to achieve this is to let arbitrarily-long sections of program be bracketed by BEGIN and END and then be seen, from outside themselves, as single statements. It is then possible to let loops, and conditional statements, be nested within each other very freely. The ALGOL-type languages allow the labelling of statements and the use of GOTO, and sometimes the best way to achieve some result is by their use. However, it does seem to be true that the use of GOTO should be deprecated since the less it is used the more elegant is the program and the more transparent its operation.

Because of its nested structure, a program in an ALGOL-type language cannot be treated as a sequence of distinct statements each sufficiently independent of its fellows to be interpreted without reference to them. The use of an ALGOL-type language therefore requires a compiler. (The word ALGOL is an abbreviation of 'Algorithmic Order Language', and successive versions of the language were devised by international committees. These have the last two digits of a year number appended, so we have ALGOL 58, ALGOL 60 and ALGOL 68. The name PASCAL is, of course, given in honour of the French mathematician of that name.)

Both types of language allow the setting-up of sections of program which can be 'called' from other parts of the program. That is to say, control can be transferred to this section by a calling statement elsewhere in the program. On completion of the operation of the section, control is returned to the statement following the calling one. In FORTRAN and BASIC these sections are called subroutines; in the ALGOL-type languages they are usually called procedures, though in PASCAL the term function is used for one which delivers a result in a particular way. The ALGOL-type procedure can be used much more flexibly than the BASIC subroutine, mainly because the former has parameters which can be assigned values when it is called. The differences will be illustrated in programs to solve the problem of the Towers of Hanoi. First it is necessary to mention some other features of PASCAL.

In PASCAL, instead of writing:

```
LET x=5
```

(as in BASIC), we use:

x:= 5

The special combination of symbols ':=' is termed the assignment operator. In PASCAL, all variables must be declared before they are used, whereas in BASIC the first use of a variable can be considered to be its declaration. The need to declare everything can be rather tedious, but can be valuable in showing-up errors in typing-in programs, since the mis-spelling of a variable name leads to a failure to compile. (In BASIC the wrongly-spelled name would be accepted as denoting a new variable, probably producing mis-leading results when the program is run.) It is essential to declare the variables in PASCAL because variables can be of many different types. In fact, the programmer is allowed to introduce new types of his own.

At the start of a PASCAL program there are normally declarations of some or all of the following, and they must be in this order:

Labels
Constants
Types
Variables
Procedures and functions

and within the declaration of each procedure or function there can be a further set of declarations, local to it. In requiring the declarations to be correctly ordered, PASCAL is somewhat more restrictive than some of its rivals - the author has to admit to a great fondness for ALGOL 68, which gives the programmer an almost embarrassing amount of freedom. (Perhaps programmers have disliked the embarrassment, for ALGOL 68 has not proved very popular.) Atkinson [2] suggests that the ordering of the declarations in PASCAL be remembered by the initial letters of:

Let's Cook Textured Vegetable Protein

A sample program illustrating the power of PASCAL is one to solve the problem of the Towers of Hanoi.

Suppose we have a set of discs, all of different diameters, and three places where we can stack them up, with the rule that no disc is ever put on top of a smaller one. Initially the discs are all stacked in one of the three locations, and of course they must be in order with the largest at the bottom of the stack and the smallest at the top. In any move a disc is taken from the top of one stack and placed on another. The problem is to move all the discs from the starting location to one of the others in as few moves as possible. It is said that monks in Hanoi are doing this with (perhaps fortunately) a substantial number of discs, making one move per day, and they believe that when the last disc is moved the world will end.

If the three locations are termed source, destination, and spare, the problem of moving n discs from source to destination, where n > 0, can be broken down as follows:

(a) Move (n-1) discs from source to spare.

(b) Move one disc from source to destination.

(c) Move (n-1) discs from spare to destination.

Of course, if n = 0, nothing at all need be done.

It follows that, if we know how to move (n-1) discs, we can move n discs. We certainly know how to move 0 discs, so we can move 1 disc and therefore we can move 2 discs, and so on. A PASCAL program to print out all the moves, for some specified number of discs, is the following:

```
PROGRAM towers(input,output);

VAR   n:integer;

PROCEDURE move(n, srce, spare, dest:integer);
BEGIN  IF n > 0 THEN
  BEGIN  move(n-1, srce, dest, spare);
         writeln('move from', srce, ' to', dest);
         move(n-1, spare, srce, dest)
  END END {move};

BEGIN  writeln('Input no. of discs');  read(n);
  move(n, 1, 2, 3)
END.
```

The first line of the program gives it the name 'towers' and indicates by the words 'input' and 'output' that it will use the standard input and output channels, namely the keyboard and screen. The main body of the program is in the last three lines, bracketed by BEGIN and END, the END being followed by a dot to confirm that this is the end of the program. What comes between the line beginning with the word PROGRAM, and the program body, is the set of declarations whose order conforms to the initial letters of 'Let's Cook Textured Vegetable Protein'. In this program there are no declarations of labels, constants or types, but there is one variable, namely n of type integer, and one procedure, namely PROCEDURE move.

The procedure is like a program within a program, and has a main body bracketed by BEGIN and END, the END being followed by the word 'move' in curly brackets so that it can be easily located. Anything in curly brackets is ignored by the compiler and is a comment intended to help the human reader, just like a line of BASIC program beginning with REM. Between the line beginning with PROCEDURE and the body of the procedure it would be possible to have a full set of declarations (in accordance with L. C. T. V. P.), applying only within the procedure. There are in fact no declarations in PROCEDURE move.

The declarations made in the main program apply also within the procedure, except when the same name is used in a local declaration, or is used to name one of the parameters of the procedure. PROCEDURE move has four parameters, all of type integer (n, srce, dest and spare). Within the procedure, n refers to the formal parameter, not to n as declared outside the procedure.

A procedure is called by giving its name. If there are parameters the name must be followed by, in brackets, a set of expressions. These will be evaluated when the procedure is called and the values assigned to the parameters. PROCEDURE move is somewhat unusual in that it contains calls of itself. These are recursive calls, and this is an example of a recursive procedure. It is important that there is a way of going through the procedure without making recursive calls, otherwise the program would never terminate once the recursive procedure had been entered. For PROCEDURE move, no recursive calls are made if it is called with n=0. The program implements the method for solution of the problem which was discussed. It can be shown that the number of moves required to transfer n discs is:

$$2^n - 1$$

The output of the 'towers' program, for four discs (and therefore 15 moves) is as follows:

```
move from    1 to    2
move from    1 to    3
move from    2 to    3
move from    1 to    2
move from    3 to    1
move from    3 to    2
move from    1 to    2
move from    1 to    3
move from    2 to    3
move from    2 to    1
move from    3 to    1
move from    2 to    3
move from    1 to    2
move from    1 to    3
move from    2 to    3
```

There are, of course, a number of differences in conventions used in PASCAL and in BASIC. In PASCAL the taking of new lines is unimportant as far as the compiler is concerned, and is used, along with indenting, to make the program presentation as clear and pleasing as possible for human readers. PASCAL employs the word 'read' (actually, a call of the system PROCEDURE read) where BASIC uses INPUT. The word READ in BASIC is kept for another purpose. Similarly, PASCAL uses PROCEDURE write or PROCEDURE writeln (the latter causing a new line to be taken after the printing) where BASIC uses PRINT.

Programs in PASCAL embody reserved words, such as PROGRAM, VAR, PROCEDURE, BEGIN, END, IF, ELSE. To make the programs comprehensible, these are normally written or printed in a way that distinguishes them from names introduced by the programmer. Here they are shown in upper-case letters, and in many texts they are shown in bold print, or sometimes underlined. The PASCAL compiler does not distinguish types of print and will always assume that these words are used for the special purposes for which they were reserved. They must not, therefore, be used in any other way (unless within a comment or text string) in a program.

The originators of the ALGOL-type languages have attached great importance to the provision of facilities for recursion. Its use in the above program makes it possible to solve a seemingly complex problem very simply. Some further reasons for regarding it as a useful facility are reviewed by Andrew [3]. Apart from the use of recursion (which is perhaps something of a red herring at this stage, since it is not widely used in the programs later in the book), the sample program may help to persuade readers that PASCAL is a language in which quite complex tasks can be presented clearly and elegantly.

TOWERS OF HANOI IN BASIC

In BASIC, it is possible to write subroutines which correspond approximately to PASCAL procedures. The section of program constituting the subroutine must contain one or more statements consisting of the word RETURN, and these have the effect of returning control to the program statement immediately following that which called the subroutine. The statement which calls a subroutine is of the form:

```
20 GOSUB 1000
```

for a subroutine beginning in line 1000.

Most implementations of BASIC allow subroutines to be called recursively. That is to say, the subroutine beginning at line 1000 may very well contain calls of itself, as:

```
1100 GOSUB 1000
```

However, the effect is not the same as that of a recursive call of a PASCAL procedure. The latter has the extra effect of storing away the values of the parameters and internal variables of the procedure, as they were just prior to the recursive call. These are held unchanged until the return is made from the recursive call, whereupon they are reinstated. The operation of the procedure, begun by the recursive call, uses a fresh set of parameters and variables, which may in turn be stored away when a further recursive call is nested in this one. The PASCAL system is doing quite a lot of work behind the scenes.

If it is a function, rather than a procedure (in the PASCAL sense in which a procedure does not return a value) that is recursive, the above explanation must be modified slightly. When the return is made from the recursive call, the value it returns is held along with the variables just reinstated. In the present application only a recursive procedure is used, so no value is returned.

PROCEDURE move has no internal variables, so it is only the parameter values which have to be stored away. They are four in number. In a BASIC program the same result can be achieved by setting up a two-dimensional array of numbers, the dimensions being twenty by four, if a depth of nesting of subroutine calls of up to twenty is to be allowed. A variable called p for 'pointer' will be used to hold the current depth of nesting and thus to point to the set of four numbers in the array which are the effective parameters of the current call. Since the sets of parameters are, in effect, stacked in the array, it is called s, for 'stack'.

A BASIC version of the above program is as follows. The DIM statement in line 20 sets the array dimensions. It can be seen that p is incremented before each recursive subroutine call, and decremented afterwards. In the INPUT statement in line 30 there is a text string which seems out of place. The INPUT statement has the facility that such a section of text is printed on the screen as a 'prompt'. For clarity, the division of program lines into lines of print is not exactly as displayed on the screen by the Sinclair SPECTRUM:

```
 10 REM Towers of Hanoi
 20 DIM s(20,4)
 30 INPUT "Enter no. of discs",n
 40 LET p=1: LET s(1,1)=n: LET s(1,2)=1:
    LET s(1,3)=2: LET s(1,4)=3
 50 REM This sets parameter values for the
    first call of move
 60 GO SUB 100
 70 STOP
100 REM subroutine move
110 IF s(p,1)=0 THEN RETURN
120 LET p=p+1: LET s(p,1)=s(p-1,1)-1:
    LET s(p,2)=s(p-1,2): LET s(p,3)=s(p-1,4):
    LET s(p,4)=s(p-1,3)
130 GO SUB 100: LET p=p-1
```

```
   140 PRINT "Move from",s(p,2)," to",s(p,4)
   150 LET p=p+1: LET s(p,1)=s(p-1,1)-1:
       LET s(p,2)=s(p-1,3): LET s(p,3)=s(p-1,2):
       LET s(p,4)=s(p-1,4)
160 GO SUB 100: LET p=p-1
170 RETURN
```

This program produces the same result as the PASCAL version, but its operation is certainly much less transparent. It is hoped that this chapter has given, to readers who did not already have it, some feeling for the characters of the two languages.

Because of its transparency, the PASCAL language has been used almost exclusively from Chapter 3 onward. A few of the programs in later Chapters use advanced features of PASCAL, but others could be translated into BASIC without much difficulty.

The Towers of Hanoi problem has received much attention in the literature. A recent paper is that of Er [4]. This treats a rather special variation of the problem, but gives references to earlier studies.

REFERENCES

1. S. Beer, **Platform for Change** (Wiley, London, 1975)

2. L. Atkinson, **Pascal Programming** (Wiley, Chichester, 1980) p. 391

3. A.M. Andrew, **Artificial Intelligence** (Abacus, Tunbridge Wells, 1983) pp. 171-172 and 176-181

4. M.C. Er, "The Colour Towers of Hanoi: A Generalization", **Computer Journal 27**, pp. 80-82 (see also correspondence on p. 90), 1984

Chapter 2

LINEAR PROGRAMMING

For mathematical optimisation the profit to be maximised (or cost to be minimised) must be expressed as a function of a number of variables, and the problem is to find what set of values assigned to these variables achieves the best result. The function is termed the Objective Function. Usually, the variables are subject to constraints - for example, the temperature of a furnace should not be allowed to go so high that its lining melts, even if the rate of production of something made in it would be maximal at the melting temperature.

There is, of course, a certain ambiguity over what is seen as a constraint and what is included in the objective function. For the furnace, it would be possible to devise an objective function representing the total profitability of the operation, taking into account the cost of repairs as a function of the temperature, and there might then be no need to take account of the constraint on temperature as such. Alternatively, a simple objective function might be used, referring only to the profit from production, and then it would be necessary to take separate account of the constraint on temperature. This point arises again in Chapter 11.

The term Linear Programming is used to refer to a set of optimisation methods which are applicable when both the objective function and the constraints are linear. They are the most frequently employed techniques of Operations Research. (Programming does not necessarily imply computer programming; the word is used as a synonym for planning.) The objective function can be written as:

$$a_1x_1 + a_2x_2 + \ldots\ldots + a_nx_n$$

where the x_i are the variables, the a_i constants, and i goes from 1 to n.

In this Chapter and the following one it will be assumed for simplicity that the Objective Function is always to be maximised. There is no loss of generality since the minimisation of a cost C can be treated as the maximisation of -C.

In the absence of constraints there is no unique optimum since, if a_1 is non-zero, we can choose x_1 to make the value of the objective function as large as we like. The problem only makes sense when there are constraints.

It is a convention of work in linear programming that variables are not allowed to be negative (though constants are). So a set of constraints which is effective, though not always explicitly stated, is:

$$x_i \geq 0, \quad \text{for all } i$$

In addition, there are linear constraints formulated by the user, say m in number. The j-th of these is of the form:

$$b_{1j}x_1 + b_{2j}x_2 + \ldots\ldots + b_{nj}x_n = c_j$$

in which the b_{ij} and c_j are constants.

The restriction of variables to non-negative values is primarily a mathematical convenience. As will be seen, the methods for solving these problems depend on it. For most practical applications of the methods the restriction is not disadvantageous. Most of the variables represent quantities which could not in fact go negative, such as the production of some commodity by a factory, or the amount of some ingredient in a feedstuff. In its application to such variables the restriction is actually necessary to ensure sensible solutions, and if it were not implicit in the method it would have to be inserted as one of the specified constraints.

Where a linear-programming problem involves a variable which is not to be restricted to non-negative values, it can be represented as the difference of two variables which are never negative, and of which at least one has value zero at any time. This is illustrated in the model-neuron example in Chapter 3, where the synaptic weights are treated in this way.

EQUATIONS AND BOUNDS

The explicitly-stated constraints are shown above as equations. However, constraints arising in practical problems are usually expressed initially as upper or lower bounds, embodying the sign $\geq$ or the sign $\leq$ rather than the equals sign. The fact that variables are restricted to non-negative values can be exploited to convert constraints imposing bounds into constraints expressed as equations and therefore more amenable to conventional mathematical treatment. This will be illustrated by a simple example.

Suppose a manufacturer has two products X1 and X2, each manufactured from three raw materials A, B and C. For each kilogram of X1 produced it is necessary to use 1 kg. of A, 2 kg. of B, and 1 kg. of C. For each kilogram of X2 produced it is necessary to use 1 kg., 1 kg., and 2 kg. of the respective ingredients. The profit per kg. of X1 is £3 and per kg. of X2 it is £2. The amounts available of A, B, and C are respectively 10, 16 and 16 kg, and the problem is to find what amounts of X1 and X2 should be made for maximum profit.

If the amounts of X1 and X2 are x_1 and x_2 respectively, the limits on availability impose the constraints:

$$x_1 + x_2 \leq 10$$

$$2x_1 + x_2 \leq 16$$

$$x_1 + 2x_2 \leq 16$$

The fact that variables are restricted to non-negative values can be exploited to convert these to equations by the introduction of further variables:

$$x_3 + x_1 + x_2 = 10$$

$$x_4 + 2x_1 + x_2 = 16$$

$$x_5 + x_1 + 2x_2 = 16$$

The new variables x_3, x_4, x_5 are termed 'slack' variables. In this example they are, of course, the unused amounts of the three raw materials. It is often possible to attach physical meanings to slack variables, but they can also be seen as the application of a mathematical trick to convert statements of bounds into equations.

The Objective Function, or representation of profit to be maximised, is denoted by B, following Vajda's [1] notation. In the present example it is given by:

$$B = 3x_1 + 2x_2$$

or, in a form similar to the constraint equations, as:

$$B - 3x_1 - 2x_2 = 0$$

In terms of the original variables x_1 and x_2, the permitted solution space is a polygon as shown in Figure 2.1. The permitted area is shown cross-hatched. Each of the lines bordering the polygon, except for the axes, corresponds to one of the user-imposed constraints. The axes, of course, correspond to the implicit constraints $x_1 \geqq 0$ and $x_2 \geqq 0$. Since the objective function is linear, it is clear that the optimum value is, in general, realised at a corner of the polygon. It can happen that there is no unique optimum because two corners give equally high values of the function, and the same value is then obtained at all points on the line joining them. It can even happen that the objective function is a constant and all points within the permitted area are equally good. However, it is always true that the optimal value is realised at one of the corners of the area.

Any assignment of values to the variables,

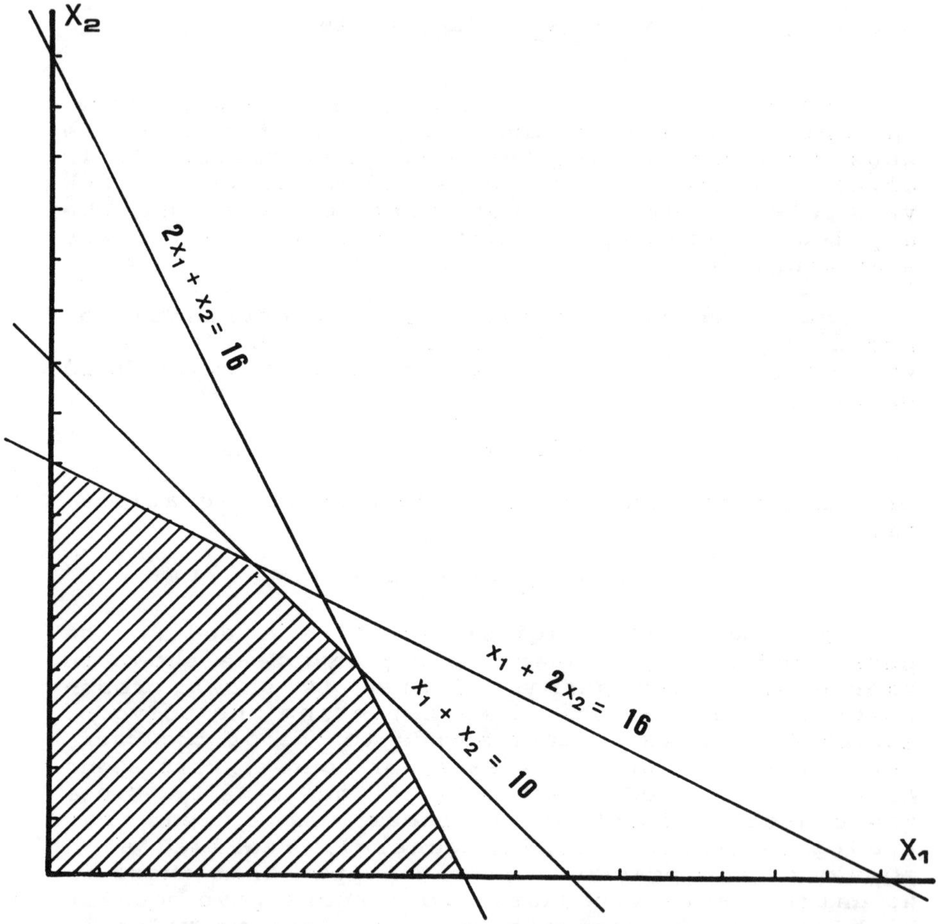

Figure 2.1. Graphical representation of problem to be solved by Linear Programming.

consistent with the constraints, is termed a Feasible Solution. Each internal point of the permitted area represents a Feasible Solution. The corners of the area represent Basic Feasible Solutions and, in view of the above, it is only necessary to examine the Basic Feasible Solutions to find an optimum. Of course, it is much easier to find the optimum by selecting it from a finite set of possibilities than from the infinity of interior points of the permitted space. Where the constraints and objective function allow the use of linear-programming methods, they are much more efficient, computationally, than certain alternatives which may have to be used if the situation is non-linear. The more general situation will be discussed in Chapter 11 when the optimisation methods of Rosenbrock and of Box are introduced.

Each boundary line of the polygon in Figure 2.1 corresponds to the assignment of a zero value to one of the five variables, including slacks, appearing in the constraint equations. Along an axis either x_1 or x_2 is zero, while along a line corresponding to a constraint the associated slack variable has value zero. Thus at a corner there are always at least two variables having value zero. At each corner in Figure 2.1 there are exactly two having value zero, but it could happen for other problems that more than two constraint lines pass through one particular corner. This would mean that more than two variables had zero values at that corner. It would also mean that one of the constraints could be removed without affecting the solution to the problem. This frequently happens in practical situations.

This analysis of a simple case illustrates (and, it is hoped, renders plausible) a mathematical definition of a Basic Feasible Solution for any number of variables and any number of constraints, provided the number of constraints is less than the number of variables. If the number of variables is n, and the number of constraints m, a Basic Feasible Solution is a Feasible Solution for which the number of variables having value zero is at least (n - m). The remaining m variables are termed basic variables and are said to form a basis.

It must be emphasised that the partitioning of the variables into a set of m basic ones and a non-basic set containing the rest, cannot be done arbitrarily. Only certain partitionings correspond to Feasible Solutions. The method to be described, termed the Simplex Method, must start from a Basic Feasible Solution, and then it moves from B.F.S. to B.F.S. until the optimum is reached. It is sometimes found that the biggest part of the solution process is the finding of the initial B.F.S. which is needed before the method can start. In Chapter 3 an extension of the Simplex Method will be discussed which allows the starting B.F.S. to be found automatically.

THE SIMPLEX METHOD

The method now to be described is completely general; any linear-programming problem can be solved using it. It is also quite simple, as its name is intended to suggest. Where the problem has certain characteristics, however, a solution is possible more economically using one of the methods to be described in Chapters 4 and 5.

The method requires the setting-up and subsequent transformation of a Simplex Tableau. This is a tabular representation of the constraint equations (in the form in which they are indeed equations), and the equation embodying the Objective Function. These equations, among them, define the problem. However, the tableau is more than a mere transcription of these equations; each presentation of a tableau denotes a particular B.F.S., and the successive transformations made to it correspond to transitions from one B.F.S. to another.

In the example considered, a B.F.S. is reached by setting both x_1 and x_2 to zero, so that these are the non-basic variables. The basic variables are then the three which were introduced as slacks. Since there are 5 variables and 3 constraints, the number of non-basic variables we must have at a B.F.S. is the difference of these, that is, two. It is obvious that this is a B.F.S. which is unlikely to be optimal, since it amounts to shutting-down the factory completely, but it allows the Simplex Method to be started.

To form the tableau it is necessary to manipulate the equations so that each basic variable is in only one equation, and has coefficient unity. The non-basic variables are allowed to appear in all the equations, and the Objective Function equation must be expressed in terms of non-basic variables only. For the example which has been considered, and the initial B.F.S. corresponding to an idle factory, the equations are already in the required forms.

Each row of the tableau represents a constraint equation, except the last one which represents the equation embodying the Objective Function. The first entry in each row is the name of the basic variable unique to it (or the profit B for the Objective-Function row). This is followed by the constant value taken from the r.h.s. of the initial equation. The remaining entries are in columns headed by the names of the non-basic variables and the entries are their coefficients in the equations. For the problem and B.F.S. considered the tableau is as follows:

		x_1	x_2
x_3	10	1	1
x_4	16	2	1
x_5	16	1	2
B	0	-3	-2

At any stage of the method, values of the variables, and of the profit, are readily deduced from the tableau. The variables at its head, here x_1 and x_2, are non-basic and have value zero. Since the equations hold with these set to zero, the values of the basic variables, and the profit, appear in the second column. Here we have $x_3 = 10$, $x_4 = 16$, $x_5 = 16$, and $B = 0$.

The numbers appearing in the lowest, or Objective Function, row, apart from the first which indicates the profit, show whether the B.F.S. represented is optimal. If it is not, they indicate how the tableau can be transformed to represent another B.F.S. corresponding to a greater profit. It can be shown that the optimum must eventually be reached by repeating such transformations. The transformation depends on exchanging one of the non-basic variables for one of the basic ones, or in other words, allowing one of the non-basic variables to take a non-zero value. Because all variables take only non-negative values, the change from a zero value must be to a positive one, and the sign of the coefficient in the bottom row shows whether this change will lead to an increase in profit. If the coefficient is negative, there will be an increase.

If all the coefficients in the bottom row are positive or zero, an optimal solution has been found. If any one of them is negative, there is the possibility of improvement. If more than one is negative, that having the lowest value (largest magnitude of negative number) is chosen to show which non-basic variable should be transferred to the basis. For the tableau as shown, the largest negative value is -3 and x_1 must join the basis.

Since the profit is increased by letting x_1 become non-zero, it should be allowed to grow to the largest value the constraints allow. It can be seen that the first constraint allows it to grow to value 10 (calculated as the constant term 10 divided by the coefficient of x_1 in the constraint equation); any further growth would require x_3 to take a negative value to balance the equation. Similarly, the second constraint limits the growth to 16/2, or 8, and the third to 16/1, or 16. (A line in which the coefficient of x_1 was zero or negative would not limit the growth at all).

It can be seen that the second constraint limits the growth of x_1 most severely, and so x_1 grows to the value of 8 determined by it. The variable x_4 then takes value zero and joins the set of non-basic variables. The column headed by the variable chosen to join the basis, here x_1, is termed the pivotal column, and the row starting with the variable which will become non-basic, here x_4, is the pivotal row. The tableau entry in which they intersect is termed the pivot.

Once the pivot has been chosen the next step is to transform the tableau according to the exchange of variables between the basic and non-basic sets. Algebraic manipulations have to be made so that the constraints, and the Objective Function equation, are expressed in terms of the new set of non-basic variables. In the case of the constraint represented by the pivotal row, all that is necessary is to divide throughout by the pivotal value, so that the new basic variable comes to have coefficient unity. Since this new basic variable, here x_1, cannot appear in any other line of the tableau, an expression for it in terms of the new set of non-basic variables has to be derived from the pivotal row, and substituted in every other row. The manipulation sounds tedious, but in fact it is equivalent to the following set of simple steps:

(a) The pivot is replaced by its reciprocal.

(b) Other values in the pivotal row are divided by the (old) pivot.

(c) Other values in the pivotal column are divided by the (old) pivot, and their sign is reversed.

(d) All other entries in the tableau have subtracted from them the amount obtained by multiplying the value in the same row and pivotal column by that in the same column and pivotal row, and dividing by the (old) pivot. The simplest way to do this in manual calculations is to visualise a rectangle having the element under consideration at one corner and the pivot at the opposite one. Then the entries at the other two corners have to be multiplied together, the product divided by the pivot and the result subtracted from the element under consideration.

Transforming the tableau in this way produces the following result:

		x_4	x_2
x_3	2	-1/2	1/2
x_1	8	1/2	1/2
x_5	8	-1/2	3/2
B	24	3/2	-1/2

For the next transformation the pivotal column is clearly the rightmost one and the pivotal row is the first since it limits the growth of x_2 to 4, while the others limit it to 16 and 16/3 respectively. Carrying out the transformation gives:

		x_4	x_3
x_2	4	-1	2
x_1	6	1	-1
x_5	2	1	-3
B	26	1	1

The positive entries in the bottom line, under x_4 and x_3, show that no further improvement is possible. The maximum possible profit of £26 is obtained when x_1, the amount of X1 manufactured, is 6 kg., and that of X2 is 4 kg. The unused amounts of raw materials A and B, represented by x_3 and x_4, are both zero since these variables are non-basic. The value of x_5 is 2, showing there are 2 kg. of raw material C unused.

COMPUTER PROGRAM

Problems to be solved by Linear Programming are usually much more complex than the above example, and a prohibitive amount of effort would be needed to solve them by hand. The method can readily be implemented as a computer program. The program which can be written at the present stage is one which is only applicable if the starting B.F.S. can be seen by inspection, and in fact the initial form of the Simplex Tableau can be inserted as data. Problems of maximisation of profit, subject to constraints imposed by availability of raw materials and/or manufacturing capacity, are often of this kind. As will be seen in the next Chapter, there are also many problems for which a starting B.F.S. is not obvious.

A program in SPECTRUM BASIC (but, again with the division into physical lines tidied-up from the presentation generated by a SPECTRUM on the screen or printer) is the following. It allows for up to ten non-basic variables, and up to ten constraints. The subscripts of the variables are held in the array b for the basic variables, and array n for the non-basic ones. The main body of the tableau is held in the array t, which has dimensions of eleven by eleven, rather than the ten by ten size which might have been expected. An extra column is provided to hold the constant values, and an extra row to hold the Objective Function row. Because the constant value column is at the left, many of the subscripts used in accessing the array t have to include an extra '+1'.

After the data have been inserted, the user is asked whether the run is to be a demonstration run. If the answer is 'Y' the tableau is displayed at each stage, but otherwise only the final tableau is shown. The subroutine starting at line 1000 displays the tableau. It makes considerable use of the comma (,) and prime (') in PRINT statements. The former causes a jump to the next TAB position and the other causes a new line to be taken. In fact, since the computer is designed to operate on low-cost peripheral devices, only two numbers are displayed per physical line and it requires a little thought to reconstruct the tableau from what is presented. A further possible source of some difficulty is that the program prints only the subscript of subscripted variables where these appear in a tableau - for example, it simply prints 5 where x_5 is intended. (This applies also to the PASCAL programs.)

The subroutine beginning in line 2000 performs the transformations of the tableau. The four rules given earlier are applied in a slightly different order, since the fourth rule is applied before numbers one and two. This is because the operation of the fourth rule requires the "old" values in the pivotal row and column.

```
 10 REM Simplex Method
 20 DIM n(10): DIM b(10): DIM t(11,11)
 30 REM array n holds nos. of non-basic
         vars., b holds nos. of basic vars.,
         t is main table of Simplex Tableau
 40 INPUT "Enter no. of non-basic vars.",nb
 50 INPUT "Enter no. of constraints",m
 60 FOR j=1 TO nb: LET n(j)=j: NEXT j
 70 FOR i=1 TO m: LET b(i)=nb+i: NEXT i
 80 PRINT "Insert tableau"
 90 FOR i=1 TO m+1: FOR j=1 TO nb+1:
    INPUT t(i, j): NEXT j: NEXT i
100 REM Initial Tableau now set up
110 INPUT "Demonstration - Y or N",d$
120 IF d$="Y" OR d$="y" THEN GO SUB 1000
130 LET min=t(m+1,2)
140 FOR j=3 TO nb+1
150 IF t(m+1,j) < min THEN
    LET min=t(m+1,j)
160 NEXT j
```

```
 170 IF min < 0 THEN GO TO 300
 180 REM Optimum has been reached
 190 IF d$ <> "Y" AND d$ <> "y" THEN
     GO SUB 1000
 200 PRINT: PRINT "Optimum reached": STOP
 300 GO SUB 2000
 310 IF d$="Y" OR d$="y" THEN GO SUB 1000
 320 GO TO 130
1000 REM Subroutine to print tableau
1010 PRINT ',,
1020 FOR j=1 TO nb: PRINT n(j),: NEXT j: PRINT
1030 FOR i=1 TO m+1: PRINT
1040 IF i <= m THEN PRINT b(i),: GO TO 1060
1050 PRINT ,
1060 FOR j=1 TO nb+1: PRINT t(i,j),: NEXT j:
     PRINT: NEXT i
1070 PRINT: RETURN
2000 REM Subroutine to transform tableau
2010 LET pcol=1
2020 FOR j=2 TO nb
2030 IF t(m+1, j+1) < t(m+1, pcol+1) THEN
     LET pcol=j
2040 NEXT j
2050 REM Pivotal column has been chosen
2060 LET prow=0
2070 FOR i=1 TO m
2080 LET x=t(i, pcol+1): IF x <= 0 THEN
     GO TO 2110
2090 LET x=t(i,1)/x: IF prow=0 THEN
     LET min=x: LET prow=i: GO TO 2110
2100 IF x < min THEN LET x=min: LET prow=i
2110 NEXT i
2120 IF prow=0 THEN
    PRINT '"Optimum at infinity": STOP
2130 REM Now to interchange variables
2140 LET x=n(pcol): LET n(pcol)=b(prow):
     LET b(prow)=x
2150 REM Now to apply the four rules
2160 LET pivot=t(prow, pcol+1):
     LET t(prow, pcol+1)=1/pivot
2170 FOR i=1 TO m+1: FOR j=1 TO nb+1:
     IF i <> prow AND j <> pcol+1 THEN
     LET t(i, j)=t(i, j)-t(prow, j)*
     t(i,pcol+1)/pivot
2180 NEXT j: NEXT i
2190 FOR j=1 TO nb+1: IF j <> pcol+1 THEN
     LET t(prow, j)=t(prow, j)/pivot
2200 NEXT j
2210 FOR i=1 TO m+1: IF i <> prow THEN
     LET t(i, pcol+1)=-t(i, pcol+1)/pivot
2220 NEXT i: RETURN
```

A program in PASCAL to do the same thing is as follows. There is an extra facility to let the final values of profit and the other variables be listed conveniently, rather than being merely displayed in the tableau.

```
PROGRAM  simplex(input,output);

VAR  m,n,i,j,pivcol,pivrow:  integer;
       {m is no. of constraints, n is no. of non-
         basic variables}
     bas:  ARRAY[1..10]  OF  integer;
     nonbas:  ARRAY[1..10]  OF  integer;
     body:  ARRAY[1..11,0..10]  OF  real;
     x,ratio:  real;   done,demo:  boolean;   ch:  char;

PROCEDURE print;
    VAR  i,j,nn:  integer;
    BEGIN  IF n>5 THEN nn:=5 ELSE nn:=n;
      write('                 ');
      FOR j:=1  TO  nn  DO  write(nonbas[j]:10);
      writeln(output);   writeln(output);
      FOR i:=1  TO  m+1  DO
        BEGIN  IF  i<=m  THEN  write(bas[i]:3,'  ')  ELSE
          write('     ');
          FOR j:=0  TO  nn  DO  write(' ',body[i,j]:9);
          writeln(output)
        END;
      writeln(output);   writeln(output)
    END {print};

PROCEDURE transform;
    VAR i,j,k:  integer;   pivot:real;
    BEGIN   k:=nonbas[pivcol];
      nonbas[pivcol]:=bas[pivrow];
      bas[pivrow]:=k;
      pivot:=body[pivrow,pivcol];
      body[pivrow,pivcol]:=1/pivot;
      FOR i:=1  TO  m+1  DO   FOR j:=0  TO  n  DO
      IF  (i<>pivrow)  AND  (j<>pivcol)  THEN
      body[i,j]:=body[i,j]-body[pivrow,j]*
        body[i,pivcol]/pivot;
      FOR i:=1  TO  m+1  DO
      IF  i<>pivrow  THEN  body[i,pivcol]:=
        -body[i,pivcol]/pivot;
      FOR j:=0  TO  n  DO
      IF  j<>pivcol  THEN  body[pivrow,j]:=
        body[pivrow,j]/pivot
    END {transform};
```

```
BEGIN  writeln('Enter no. of non-basic variables');
  read(n);
  writeln('Enter no. of constraints');  readln(m);
  writeln('Print all steps?  Enter y or n');
  read(ch);  demo:=(ch='Y') OR (ch='y');

  FOR j:=1 TO n DO nonbas[j]:=j;
  FOR i:=1 TO m DO bas[i]:=n+i;

  writeln(
    'Enter body of simplex tableau, row by row,');
  writeln('each row beginning with const. term.');
  FOR i:=1 TO m+1 DO  FOR j:=0 TO n DO
    read(body[i,j]);

  IF demo THEN print;

  REPEAT  x:=body[m+1,1];  pivcol:=1;
    FOR j:=2 TO n DO
    IF body[m+1,j]<x THEN
    BEGIN  x:=body[m+1,j];  pivcol:=j  END;
    done:=x>=0;
    IF NOT done THEN
    BEGIN  pivrow:=0;
      FOR i:=1 TO m DO
      IF body[i,pivcol]>0 THEN
      BEGIN  ratio:=body[i,0]/body[i,pivcol];
        IF (pivrow=0) OR (ratio<x) THEN
        BEGIN  x:=ratio;  pivrow:=i  END
      END;
      IF pivrow>0 THEN
      BEGIN  transform;  IF demo THEN print  END;
    END;
  UNTIL done OR (pivrow=0);

  IF pivrow=0 THEN writeln('Optimum at infinity')
  ELSE BEGIN writeln('Optimum found');
    writeln(output);
    writeln('Profit = ', body[m+1,0]);
    writeln(output);
    writeln('Values as follows:');  writeln(output);
    writeln('Var. no.      Value');  writeln(output);
    FOR j:=1 TO n DO  writeln(nonbas[j]:6,'  ',0);
    FOR i:=1 TO m DO  writeln(bas[i]:6,'  ',
      body[i,0])
  END
END.
```

Where numerical values are included in write and writeln statements, they are, in this program, followed by a colon and an extra number. This is a feature of PASCAL not previously used, and it allows control of the number of printing spaces used when the number is printed. This is a great help in achieving a pleasing layout of results, especially when numbers have to be arranged in tables. It can be seen that PROCEDURE print is made to print only the first five headed columns of the Tableau. This avoids over-running the line length, and is adequate for simple demonstration programs. The simple example discussed requires only two headed columns. The print-out of the program, in its demonstration mode, is as follows. :

```
                        1          2

 3    1.00E+01  1.00E+00  1.00E+00
 4    1.60E+01  2.00E+00  1.00E+00
 5    1.60E+01  1.00E+00  2.00E+00
      0.00E+00 -3.00E+00 -2.00E+00


                        4          2

 3    2.00E+00 -5.00E-01  5.00E-01
 1    8.00E+00  5.00E-01  5.00E-01
 5    8.00E+00 -5.00E-01  1.50E+00
      2.40E+01  1.50E+00 -5.00E-01


                        4          3

 2    4.00E+00 -1.00E+00  2.00E+00
 1    6.00E+00  1.00E+00 -1.00E+00
 5    2.00E+00  1.00E+00 -3.00E+00
      2.60E+01  1.00E+00  1.00E+00

Optimum found

Profit =  2.6000000E+01

Values as follows:

Var. no.     Value

     4             0
     3             0
     2    4.0000000E+00
     1    6.0000000E+00
     5    2.0000000E+00
```

MATHEMATICAL PROGRAMMING

Linear Programming is the simplest case of a more general area of study in which problems of optimisation are solved by computationally-efficient methods. The term Mathematical Programming is sometimes used to refer to this more general field, where the Objective Function and constraints are not necessarily of linear form. Boot [2], for example, has presented methods for a certain range of problems under the heading of Quadratic Programming.

The literature on Linear Programming is large and includes a number of very good textbooks. A small one which introduces the ideas very clearly, and which has influenced the author and hence the method of presentation here, is by Vajda [1]. Others are listed in the suggestions for Further Reading.

REFERENCES

1. S. Vajda, **An Introduction to Linear Programming and the Theory of Games** (Methuen, London 1960 and 1963, and in 1966, 1970 and 1978 in Methuen Science Paperback Series)

2. J.C.G. Boot, **Quadratic Programming** (North-Holland, Amsterdam, 1964)

Chapter 3

FINDING THE INITIAL BASIC FEASIBLE SOLUTION

THE BIG-M METHOD

The great shortcoming of the Simplex Method, in the simple form in which it was introduced in the last Chapter, is the need for the user to find a Basic Feasible Solution from which the process can be started. In situations like the one which was used for illustration, a starting B.F.S. can be found without difficulty, but this is not so in general. An L.P. program for general use would not be very satisfactory if it required its users to perform a large amount of analysis and preliminary manipulation of the problem formulation.

Fortunately, a means has been found of automating the finding of a starting B.F.S. In fact, it turns out that it can be made to depend on exactly the same kind of optimisation as does the main problem. The solution process therefore proceeds in two similar stages - a first in which the starting B.F.S. is found, and the second to find the optimum.

As was seen in Chapter 2, there is no difficulty when the comparator in the constraint formulation is 'less than or equals', or $\leq$. For constraints using this comparator, the slack variable which is introduced can become part of the basis and is already conveniently present with coefficient unity.

Where the comparator is 'greater than or equals', or $\geq$, the slack variable has coefficient -1, so is unsuitable. It can also happen that some constraints are given as equations, and then there is no slack variable at all. (It might seem that, where a constraint is presented as an equation, algebraic manipulation could be used to eliminate this constraint as well as one of the variables. However, it is safer to allow the constraint to remain because the elimination of a variable also removes the implicit constraint arising from the restriction of this variable to non-negative values.)

It is possible to change a greater-than constraint into a less-than one by multiplying both sides by -1 and changing the comparator. This, however, is not permissible - the Simplex Method depends on having non-negative values for the constant terms, since these values have to be assigned to variables forming the current basis. (When carrying out the method by hand, it is a sign that something has gone wrong if at some stage a negative value appears in the constant-term column.)

The trick which is used seems at first sight to be preposterous - if a constraint equation does not contain a variable suitable to become an element of the basis, a new variable is simply written-in. Then the first stage of the optimisation process is to bring these 'artificial variables' to zero, and in fact to make them all non-basic. Once this has been done the artificial variables can be removed from the tableau, which must now have a basis formed of other variables. It therefore represents a Basic Feasible Solution for the original problem.

Linear Programming is sometimes applied to the mixing of feedstuffs, or of, say, fertilisers, so as to achieve a certain specification at minimum cost. The optimal mix depends on the costs of the available raw materials, and on the content of each of the significant ingredients in each of them.

Suppose, for example, that a farmer wants to apply minimum amounts of three nutrients as follows:

Potassium	(as K_2O)	5 kg.
Nitrogen	(as N)	3 kg.
Phosphorus	(as P_2O_5)	4 kg.

Three fertilisers, A, B and C, are available, and their details are as follows:

	$\%K_2O$	$\%N$	$\%P_2O_5$	Cost/kg.
A	10	12	3	60
B	2	8	10	40
C	6	5	4	40

Since the constraints refer to minimum amounts of the three important nutrients, it is clear that they are violated if no fertiliser at all is applied. In contrast to the previous example, setting all the principal variables to zero does not give a Basic Feasible Solution. In this simple example it is not too difficult to find a B.F.S. in which all the requirements are met by applying only one of the three kinds of fertiliser, but this entails the kind of analysis and algebraic manipulation which should not be required of a user. The constraints imposed by the minimum requirements for nutrients are as follows, where the amounts of the fertilisers are respectively x_1, x_2 and x_3:

$$\begin{aligned} 10x_1 + 2x_2 + 6x_3 &\geq 500 \\ 12x_1 + 8x_2 + 5x_3 &\geq 300 \\ 3x_1 + 10x_2 + 4x_3 &\geq 400 \end{aligned}$$

In order to turn the constraints into equations, slacks are introduced as before, but this time with the minus sign:

$$\begin{aligned} 10x_1 + 2x_2 + 6x_3 - x_4 &= 500 \\ 12x_1 + 8x_2 + 5x_3 - x_5 &= 300 \\ 3x_1 + 10x_2 + 4x_3 - x_6 &= 400 \end{aligned}$$

Since none of the slack variables has coefficient unity, none of them can be included in the basis, and artificial variables have to be introduced in all three lines. It is a convention that subscripts in excess of 100 are given to these, so that they can be easily distinguished.

$$10x_1 + 2x_2 + 6x_3 - x_4 + x_{101} = 500$$

$$12x_1 + 8x_2 + 5x_3 - x_5 + x_{102} = 300$$

$$3x_1 + 10x_2 + 4x_3 - x_6 + x_{103} = 400$$

The first stage of the optimisation is to bring each of x_{101}, x_{102} and x_{103} to zero, if this is possible. It can happen that it is impossible to bring the artificial variables to zero, and it then follows that there is no Feasible Solution to the original problem. Because the artificial variables are like other variables in Linear Programming, in that they cannot go negative, all three of them have the value zero if, and only if, their sum has value zero. Since the sum cannot go negative, bringing it to zero is the same as minimising it. The first stage of the optimisation process, then, is one in which the function to be minimised is:

$$x_{101} + x_{102} + x_{103}$$

or, equivalently, what is to be maximised is:

$$-(x_{101} + x_{102} + x_{103})$$

However, this is not a suitable form for an Objective Function, since it is a requirement of the method that only non-basic variables appear in this function. The artificial variables are basic. From the constraint equations it is easy to express each of the artificial variables in terms of the other (non-basic) ones, and these expressions have to be substituted in the above Objective Function. We have for example, for x_{101}:

$$x_{101} = -10x_1 - 2x_2 - 6x_3 + x_4 + 500$$

Substituting in the Objective Function gives:

$$B = -25x_1 - 20x_2 - 15x_3 + x_4 + x_5 + x_6 + 1200$$

or

$$B + 25x_1 + 20x_2 + 15x_3 - x_4 - x_5 - x_6 = 1200$$

It can be seen that the Objective Function row which is effective during the first stage of optimisation is formed by adding together the entries in those rows which contain artificial variables.

Once all the artificial variables have been brought to zero, the second stage of optimisation is entered, and here the Objective Function is the appropriate one for the problem as first presented. The cost of fertiliser applied is:

$$60x_1 + 40x_2 + 40x_3$$

and so the function to be maximised is:

$$-(60x_1 + 40x_2 + 40x_3)$$

represented in an equation as follows:

$$B + 60x_1 + 40x_2 + 40x_3 = 0$$

The combination of the two stages may be thought of as a single optimisation in which the Objective Function is a composite one obtained by multiplying that for the first stage by a number M, and adding the product to the Objective Function for the second stage. The number M is never evaluated; it is assumed to be so large that, as long as the first-stage Objective Function is non-zero, it dominates the other.

Because of the appearance of M the method is sometimes given the picturesque name of 'Big-M Method'.

In the working programs which will be shown, M is used only notionally, and in fact the switch-over from one Objective Function to the other is determined rather differently from what the foregoing would suggest. The difficulty about using a composite Objective Function is that the values in the first-stage part of it may not come precisely to zero when the first stage is finished, because of rounding-off errors in the computation. In the program an artificial variable is removed as soon as it becomes non-basic. Since it is non-basic it heads a column of the tableau, and the whole column is removed and the tableau closed up.

If there are still artificial variables in the tableau when it is found that no further improvement is possible in the first stage, the method must terminate with the message "No Solution Possible".

If it proves possible to eliminate all the artificial variables, the method then switches over to the second-stage Objective Function, which is that appropriate to the problem as originally presented.

In this example all three of the constraints have the comparator $\geq$. There is no reason why constraints of different kinds should not appear in the same problem. Constraints having the comparator $\leq$, do not require an artificial variable, since the slack variable can enter the basis. Constraints having the equals sign do not require the addition of a slack variable, and an artificial variable has to be inserted exactly as for constraints of the 'greater than or equals' kind. The following is a PASCAL program allowing the two-stage operation of the Big-M Method:

```
PROGRAM bigm(input,output);

VAR  m,n,nb,i,j,pivcol,pivrow,stage,avn: integer;
        {m is no. of constraints, n is current no. of
         non-basic variables, nb the initial no.
         specifiedby user}
bas: ARRAY[1..10] OF integer;
nonbas: ARRAY[1..20] OF integer;
body: ARRAY[1..12,0..20] OF real;
x,ratio: real;  done,impos,demo: boolean;  ch: char;

PROCEDURE print;
    VAR  i,j,nn: integer;
    BEGIN  IF n>5 THEN nn:=5 ELSE nn:=n;
      write('              ');
      FOR j:=1 TO nn DO write(nonbas[j]:10);
      writeln(output);  writeln(output);
      FOR i:=1 TO m+3-stage DO
        BEGIN  IF i<=m THEN write(bas[i]:4,' ') ELSE
          write('     ');
          FOR j:=0 TO nn DO write(' ',body[i,j]:9);
          writeln(output)
        END;
      writeln(output);  writeln(output)
    END {print};

PROCEDURE transform;
    VAR i,j,k: integer;  pivot:real;
    BEGIN  k:=nonbas[pivcol];  nonbas[pivcol]:=
      bas[pivrow];
      bas[pivrow]:=k;
      pivot:=body[pivrow,pivcol];
```

```
      body[pivrow,pivcol]:=1/pivot;
      FOR i:=1 TO m+3-stage DO  FOR j:=0 TO n DO
      IF (i<>pivrow) AND (j<>pivcol) THEN
      body[i,j]:=body[i,j]-body[pivrow,j]*
        body[i,pivcol]/pivot;
      FOR i:=1 TO m+3-stage DO
      IF i<>pivrow THEN body[i,pivcol]:=
        -body[i,pivcol]/pivot;
      FOR j:=0 TO n DO
      IF j<>pivcol THEN body[pivrow,j]:=
        body[pivrow,j]/pivot
    END {transform};

PROCEDURE remove;
    VAR  i,j: integer;
    BEGIN  n:=n-1;
      FOR j:=pivcol TO n DO
      BEGIN  nonbas[j]:=nonbas[j+1];
        FOR i:=1 TO m+2 DO  body[i,j]:=body[i,j+1]
      END;
      stage:=2;
      FOR i:=1 TO m DO  IF bas[i]>100 THEN stage:=1;
      IF (stage=2) AND demo THEN print
    END {remove};

BEGIN  writeln('Enter no. of non-basic variables');
  read(nb);
  writeln('Enter no. of constraints');  readln(m);
  writeln('Print all steps?  Enter y or n');
  read(ch);  demo:=(ch='Y') OR (ch='y');  n:=nb;

  FOR j:=1 TO n DO nonbas[j]:=j;
  FOR i:=1 TO m DO bas[i]:=n+i;
  avn:=100;

  writeln(
    'Enter body of simplex tableau, row by row,');
  writeln('each row beginning with const. term.');
  FOR i:=1 TO m DO
  BEGIN  FOR j:=0 TO nb DO  read(body[i,j]);
    FOR j:=nb+1 TO 20 DO body[i,j]:=0;
    writeln('Lower bound, upper, or exact value?');
    writeln('Enter l, u or e');
    readln(input);  read(ch);
    IF (ch='l') OR (ch='L') THEN
      BEGIN  n:=n+1;  avn:=avn+1;  nonbas[n]:=bas[i];
        bas[i]:=avn;  body[i,n]:=-1  END;
    IF (ch='e') OR (ch='E') THEN
      BEGIN  avn:=avn+1;  bas[i]:=avn  END
  END;
```

```
  FOR j:=0 TO nb DO read(body[m+1,j]);
  FOR j:=nb+1 TO 20 DO body[m+1,j]:=0;
  FOR j:=0 TO 20 DO body[m+2,j]:=0;
  FOR i:=1 TO m DO
    IF bas[i]>100 THEN FOR j:=0 TO n DO
      body[m+2,j]:=body[m+2,j]-body[i,j];

  stage:=1;
  IF demo THEN print;

  REPEAT  x:=body[m+3-stage,1];  pivcol:=1;
    FOR j:=2 TO n DO
    IF body[m+3-stage,j]<x THEN
    BEGIN  x:=body[m+3-stage,j];  pivcol:=j  END;
    IF x>=0 THEN
      BEGIN  done:=true;
        IF stage=1 THEN impos:=true  END;
    IF NOT done THEN
    BEGIN  pivrow:=0;
      FOR i:=1 TO m DO
      IF body[i,pivcol]>0 THEN
      BEGIN  ratio:=body[i,0]/body[i,pivcol];
        IF (pivrow=0) OR (ratio<x) THEN
        BEGIN  x:=ratio;  pivrow:=i  END
      END;
      IF pivrow>0 THEN
      BEGIN  transform;  IF demo THEN print;
        IF nonbas[pivcol]>100 THEN remove  END;
    END;
  UNTIL done OR (pivrow=0);

  IF pivrow=0 THEN writeln('Optimum at infinity')
  ELSE IF impos THEN
    writeln('Constraints incompatible')
  ELSE BEGIN
    writeln('Optimum found');  writeln(output);
    writeln('Profit = ',body[m+1,0]);
    writeln(output);
    writeln('Values as follows:');  writeln(output);
    writeln('Var. no.     Value');  writeln(output);
    FOR j:=1 TO n DO  writeln(nonbas[j]:6,'  ',0);
    FOR i:=1 TO m DO
      writeln(bas[i]:6,'  ',body[i,0])
  END
END.
```

The user is asked to insert an initial number of non-basic variables, stored within the program as nb. This will be the number of variables in the original specification of the problem. As constraints are introduced, further non-basic variables may be needed, and these are added by the program. The value n is initially set equal to nb but is then incremented as required, whenever it is necessary to introduce a slack variable which does not join the basis. The integer 'avn' (for Artificial Variable Number) is initially set to 100, and then incremented and used to provide the subscript number, every time an artificial variable has to be introduced.

Whenever an artificial variable becomes non-basic, PROCEDURE remove is called to decrement n and to remove the column (the pivotal column in the last transformation of the tableau) headed by the variable. It closes-up the tableau, and checks if there are any artificial variables left in it. If there are none, the procedure sets the variable 'stage' to 2.

The print-out of this program, using the fertiliser example, is as follows. To show all the columns, the print-outs of the tableau, except the last, have been split vertically and the right-hand part moved down the page. (Since there is therefore no difficulty in printing all the columns, the program was modified so that PROCEDURE print did not limit the number of headed columns to five.)

		1	2	3
101	5.00E+02	1.00E+01	2.00E+00	6.00E+00
102	3.00E+02	1.20E+01	8.00E+00	5.00E+00
103	4.00E+02	3.00E+00	1.00E+01	4.00E+00
	0.00E+00	6.00E+01	4.00E+01	4.00E+01
	-1.20E+03	-2.50E+01	-2.00E+01	-1.50E+01

4	5	6
-1.00E+00	0.00E+00	0.00E+00
0.00E+00	-1.00E+00	0.00E+00
0.00E+00	0.00E+00	-1.00E+00
0.00E+00	0.00E+00	0.00E+00
1.00E+00	1.00E+00	1.00E+00

		102	2	3
101	2.50E+02	-8.33E-01	-4.67E+00	1.83E+00
1	2.50E+01	8.33E-02	6.67E-01	4.17E-01
103	3.25E+02	-2.50E-01	8.00E+00	2.75E+00
	-1.50E+03	-5.00E+00	0.00E+00	1.50E+01
	-5.75E+02	2.08E+00	-3.33E+00	-4.58E+00

4	5	6
-1.00E+00	8.33E-01	0.00E+00
0.00E+00	-8.33E-02	0.00E+00
0.00E+00	2.50E-01	-1.00E+00
0.00E+00	5.00E+00	0.00E+00
1.00E+00	-1.08E+00	1.00E+00

		2	1	4
101	1.40E+02	-7.60E+00	-4.40E+00	-1.00E+00
3	6.00E+01	1.60E+00	2.40E+00	0.00E+00
103	1.60E+02	3.60E+00	-6.60E+00	0.00E+00
	-2.40E+03	-2.40E+01	-3.60E+01	0.00E+00
	-3.00E+02	4.00E+00	1.10E+01	1.00E+00

5	6
1.20E+00	0.00E+00
-2.00E-01	0.00E+00
8.00E-01	-1.00E+00
8.00E+00	0.00E+00
-2.00E+00	1.00E+00

		2	1	4
5	1.17E+02	-6.33E+00	-3.67E+00	-8.33E-01
3	8.33E+01	3.33E-01	1.67E+00	-1.67E-01
103	6.67E+01	8.67E+00	-3.67E+00	6.67E-01
	-3.33E+03	2.67E+01	-6.67E+00	6.67E+00
	-6.67E+01	-8.67E+00	3.67E+00	-6.67E-01

```
        101          6

     8.33E-01  0.00E+00
     1.67E-01  0.00E+00
    -6.67E-01 -1.00E+00
    -6.67E+00  0.00E+00
     1.67E+00  1.00E+00


                        103          1          4

 5   1.65E+02  7.31E-01 -6.35E+00 -3.46E-01
 3   8.08E+01 -3.85E-02  1.81E+00 -1.92E-01
 2   7.69E+00  1.15E-01 -4.23E-01  7.69E-02
    -3.54E+03 -3.08E+00  4.62E+00  4.62E+00
    -7.63E-06  1.00E+00 -2.15E-06  2.38E-07

          6

    -7.31E-01
     3.85E-02
    -1.15E-01
     3.08E+00
    -1.19E-07


                          1          4          6

 5   1.65E+02 -6.35E+00 -3.46E-01 -7.31E-01
 3   8.08E+01  1.81E+00 -1.92E-01  3.85E-02
 2   7.69E+00 -4.23E-01  7.69E-02 -1.15E-01
    -3.54E+03  4.62E+00  4.62E+00  3.08E+00

Optimum found

Profit = -3.5384617E+03

Values as follows:

Var. no.     Value

     1            0
     4            0
     6            0
     5   1.6538462E+02
     3   8.0769235E+01
     2   7.6923078E+00
```

The print-outs of the tableau made in stage one show two rows embodying the Objective Function. (In this example all except the last print-out belong to the first stage, because the first tableau after elimination of the artificial variables happens to represent the optimal solution.) It is to be understood that the effective Objective Function row is the sum of (a) the upper of the two rows and (b) the lower one multiplied by M.

This program still requires the user to devise the main part of the tableau, even though it attends automatically to the introduction of slack and artificial variables. Where the application is like the last example, it is likely that all the constraints refer either to amounts of an ingredient (like Fertiliser A, B or C) or of a constituent (like Potassium, Nitrogen or Phosphorus). A modification of the program which could be more convenient when applied to the mixing of feedstuffs or fertilisers is given below.

Constraints which are likely to enter into mixing problems can be neatly represented by specifying four things:

(a) The critical numerical value which will become the constant term on the right-hand side of the constraint.

(b) An indication whether the constraint refers to an ingredient or a constituent.

(c) An integer indicating which ingredient or which constituent.

(d) An indication of the relationship as 'greater than or equals', 'less than or equals', or 'equals'.

In the following program, a variable-type with the name 'cnstr' has been declared as a RECORD type, with four fields suitable for holding these items of information. Two of the fields, namely those named 'reference' and 'relation', hold variables of a new kind, having a set of alternative values listed in the type-declaration. The field 'relation' contains a variable which can take either of the two values 'ingredient' and 'constituent', and similarly that in 'relation' can only hold 'upper', 'lower' or 'equal'. The set of constraints is held as an array of variables of type 'cnstr'.

One way of referring to a particular field is by specifying the record containing it, as for instance by writing 'constraints[i]', and following this by a dot and the name of the desired field. Thus the 'reference' field of 'constraints[i]' can be indicated by writing 'constraints[i].reference'.

However, it is often more convenient to use the WITH .. DO construction. A program statement (or group of statements bracketed by BEGIN and END), can be preceded by, say, 'WITH constraints[i] DO'. The effect is that, within this statement or group of statements it is understood that 'reference' is equivalent to 'constraints[i].reference', and so on. In the Feedmix program which follows, references to fields of the records representing constraints are made using the WITH .. DO construction.

```
PROGRAM feedmix(input,output);

TYPE cnstr= RECORD    value: real;
                      reference: (ingredient,
                                  constituent);
                      no: integer;
                      relation: (upper,lower,equal)
END;

VAR  m,n,nb,nc,i,j,pivcol,pivrow,stage,avn: integer;
        {m is no. of constraints, n is current no. of
        non-basic variables, nb the initial no. of
        variables specified by user, in this case
        equal to the number of permissible
        ingredients, and nc is the number of constit-
        uents specified}
     bas: ARRAY[1..10] OF integer;
     nonbas: ARRAY[1..20] OF integer;
     body: ARRAY[1..12,0..20] OF real;
     analysis: ARRAY[1..10,1..10] OF real;
     price: ARRAY[1..10] OF real;
     constraints: ARRAY[1..10] OF cnstr;
     x,ratio: real;  done,impos,demo: boolean;
     ch:char;

PROCEDURE print;
    VAR  i,j,nn: integer;
    BEGIN  IF n>5 THEN nn:=5 ELSE nn:=n;
      write('              ');
      FOR j:=1 TO nn DO write(nonbas[j]:10);
```

```
      writeln(output);   writeln(output);
      FOR i:=1 TO m+3-stage DO
        BEGIN  IF i<=m THEN write(bas[i]:4,' ') ELSE
          write('     ');
          FOR j:=0 TO nn DO write(' ',body[i,j]:9);
          writeln(output)
        END;
      writeln(output);   writeln(output)
    END {print};

PROCEDURE transform;
    VAR i,j,k: integer;  pivot:real;
    BEGIN  k:=nonbas[pivcol];  nonbas[pivcol]:=
      bas[pivrow];
      bas[pivrow]:=k;
      pivot:=body[pivrow,pivcol];
      body[pivrow,pivcol]:=1/pivot;
      FOR i:=1 TO m+3-stage DO  FOR j:=0 TO n DO
      IF (i<>pivrow) AND (j<>pivcol) THEN
      body[i,j]:=body[i,j]-
        body[pivrow,j]*body[i,pivcol]/pivot;
      FOR i:=1 TO m+3-stage DO
      IF i<>pivrow THEN body[i,pivcol]:=
        -body[i,pivcol]/pivot;
      FOR j:=0 TO n DO
      IF j<>pivcol THEN body[pivrow,j]:=
        body[pivrow,j]/pivot
    END {transform};

PROCEDURE remove;
    VAR  i,j: integer;
    BEGIN  n:=n-1;
      FOR j:=pivcol TO n DO
      BEGIN  nonbas[j]:=nonbas[j+1];
        FOR i:=1 TO m+2 DO  body[i,j]:=body[i,j+1]
      END;
      stage:=2;
      FOR i:=1 TO m DO  IF bas[i]>100 THEN stage:=1;
      IF (stage=2) AND demo THEN print
    END {remove};

BEGIN  writeln('Enter no. of ingredients');
  read(nb);
  writeln('Enter no. of constituents or properties');
  read(nc);
  writeln('Enter analysis table');
  FOR i:=1 TO nb DO  FOR j:=1 TO nc DO
    read(analysis[i,j]);
  writeln('Enter prices');
  FOR i:=1 TO nb DO read(price[i]);
```

```
writeln('Enter no. of constraints');  readln(m);
FOR i:=1 TO m DO
  WITH constraints[i] DO
    BEGIN  writeln('Constraint no. ',i:2);
      writeln('Enter critical value');
      readln(value);
      writeln(
        'Ingredient or constituent?  Enter i or c');
      read(ch);
      IF (ch='i') OR (ch='I') THEN reference:=
     ingredient
           ELSE reference:=constituent;
      writeln('Which number?');   readln(no);
      writeln('Upper bound, lower, or equality?');
      writeln('Enter u, l or e');  readln(ch);
      IF (ch='u') OR (ch='U') THEN relation:=
        upper
      ELSE IF (ch='l') OR (ch='L') THEN relation:=
        lower
      ELSE relation:= equal
    END;

writeln('Print all steps?  Enter y or n');
read(ch);  demo:=(ch='Y') OR (ch='y');  n:=nb;

FOR j:=1 TO n DO nonbas[j]:=j;
FOR i:=1 TO m DO bas[i]:=n+i;
avn:=100;

FOR i:=1 TO m DO
WITH constraints[i] DO
BEGIN  body[i,0]:=value;
  FOR j:=1 TO nb DO
  IF reference=ingredient THEN
  BEGIN  IF  j=no THEN body[i,j]:=1
    ELSE body[i,j]:=0  END
    ELSE  body[i,j]:=analysis[j,no];
    FOR j:=nb+1 TO 20 DO body[i,j]:=0;
    IF relation=lower THEN
    BEGIN  n:=n+1;  avn:=avn+1;  nonbas[n]:=bas[i];
      bas[i]:=avn;  body[i,n]:=-1  END;
    IF relation=equal THEN
    BEGIN  avn:=avn+1;  bas[i]:=avn  END
  END;

FOR j:=0 TO n DO body[m+1,j]:=0;
FOR j:=1 TO nb DO body[m+1,j]:=price[j];
FOR j:=0 TO n DO body[m+2,j]:=0;
FOR i:=1 TO m DO
  IF bas[i]>100 THEN FOR j:=0 TO n DO
    body[m+2,j]:=body[m+2,j]-body[i,j];
```

```
    stage:=1;
    IF demo THEN print;

  REPEAT  x:=body[m+3-stage,1];  pivcol:=1;
    FOR j:=2 TO n DO
    IF body[m+3-stage,j]<x THEN
    BEGIN  x:=body[m+3-stage,j];  pivcol:=j  END;
    IF x>=0 THEN
      BEGIN  done:=true;
        IF stage=1 THEN impos:=true  END;
    IF NOT done THEN
    BEGIN  pivrow:=0;
      FOR i:=1 TO m DO
      IF body[i,pivcol]>0 THEN
      BEGIN  ratio:=body[i,0]/body[i,pivcol];
        IF (pivrow=0) OR (ratio<x) THEN
        BEGIN  x:=ratio;  pivrow:=i  END
      END;
      IF pivrow>0 THEN
      BEGIN  transform;  IF demo THEN print;
        IF nonbas[pivcol]>100 THEN remove  END;
    END;
  UNTIL done OR (pivrow=0);

  IF pivrow=0 THEN writeln('Optimum at infinity')
  ELSE IF impos THEN
    writeln('Constraints incompatible')
  ELSE BEGIN
    writeln('Optimum found');  writeln(output);
    writeln('Profit = ',body[m+1,0]);
    writeln(output);
    writeln('Values as follows:');  writeln(output);
    writeln('Var. no.     Value');  writeln(output);
    FOR j:=1 TO n DO  writeln(nonbas[j]:6,'  ',0);
    FOR i:=1 TO m DO
      writeln(bas[i]:6,'  ',body[i,0])
  END
END.
```

Apart from the way constraints and prices are represented, this program is very similar to the earlier Big-M one. For the example considered earlier, the analysis table to be inserted would be as follows:

0.1	0.12	0.03
0.02	0.08	0.1
0.06	0.05	0.04

being a tabular representation of the potassium,

nitrogen and phosphorus content of fertilisers A, B and C. In this problem there was no prior specification of the total quantity of fertiliser to be used. In formulating a feedstuff, or a fertiliser to be sold in bags to farmers, it might be necessary to provide the required constituents in a specified weight, say 10 kg., of the mix. This could be arranged by having a further row in the analysis table referring to weight as though it were a constituent, and introducing a constraint treating weight as a constituent required to have the exact value of 10 units. The extra row of the table would simply be:

1.0 1.0 1.0

but would be filled with reciprocals of densities if the total amount of mix was specified as a volume rather than a weight.

In practice it would be convenient to have the price data, analysis, and list of constraints, held in arrays as in this program. In a program for practical use they would no doubt be stored in computer files between runs, and there would be means of editing them, so that values could be updated without the need to re-enter all the data from the beginning. The programs in this book are all capable of this sort of improvement; it should be possible to enter data and to store it in files and edit it, both for updating and for correction of errors. For simplicity the programs have been written such that the data must be entered manually for each run. In this form the programs are perfectly suitable for running simple illustrative examples. For serious use they should be modified in ways which are quite straightforward but would be tedious and confusing here.

USING STAGE ONE ONLY

Sometimes it is useful to employ only the first stage of the Big-M Method to test whether any Feasible Solution exists for a set of constraints. An example arises in connection with the Model Neurons discussed by various workers, the first being McCulloch and Pitts [1]. A model neuron is essentially a threshold element; a number of inputs to it constribute to a quantity summed within it, and if the summation exceeds a threshold the neuron fires to produce an output, and otherwise it does not. The contribution made by one input, if active, is termed its 'weight'. It is not an entirely trivial matter to decide whether there is an allocation of weight and threshold values which will allow a model neuron to compute some given logical function of its inputs. One way of answering the question is by an adaptation of Linear Programming.

The weight values can be either positive or negative. In neurophysiological terms, an input, or synapse, is called excitatory if its weight is positive, and inhibitory if its weight is negative. Strictly speaking, threshold elements of this kind are not what McCulloch and Pitts described, since they stipulated inhibitory synapses which, when active, imposed a total veto on firing. However, the threshold-element model is mathematically tractable and has been preferred by most theoretical neurophysiologists following the general lead of McCulloch and Pitts. In particular, threshold devices are assumed throughout the extensive literature on the 'Perceptron' principle for pattern recognition, very clearly discussed by Nilsson [2].

Suppose we wish to decide whether a neuron with three inputs can be set to fire when its inputs are 1, 1, 0 or when they are 1, 0, 1, but not when they are 1, 1, 1. The weights of the three inputs can be called w_1, w_2 and w_3. Since the weights can take either sign, it is necessary to express each of them as the difference of two variables, these variables being always non-negative. The same goes for the threshold, t. We can make the substitutions:

$$w_1 = x_1 - x_2 \qquad w_2 = x_3 - x_4$$

$$w_3 = x_5 - x_6 \qquad t = x_8 - x_7$$

To write the constraints it is convenient to suppose that the summations 'miss' the threshold on the appropriate side by a finite margin, say one unit. The constraints imposed by the requirements can then be written:

$$\begin{aligned} w_1 + w_2 &\geqq t + 1 \\ w_1 + w_3 &\geqq t + 1 \\ w_1 + w_2 + w_3 &\leqq t - 1 \end{aligned}$$

On making substitutions of x-values for w-values, and reversing the sign of the third constraint, it is found that each constraint requires a slack variable having negative sign and therefore also requires an artificial variable. With slacks x_9, x_{10}, x_{11}, and artificial variables x_{101}, x_{102}, x_{103}, the constraints can be formed into a Simplex Tableau as follows. The Objective Function row is the one which would normally be multiplied by M, as needed to try to bring the artificial variables to zero. It is formed as the negative of the sum of coefficients in those rows (here, all of them) which include artificial variables.

		x_1	x_2	x_3	x_4	x_5	x_6	x_7	x_8	x_9	x_{10}	x_{11}
x_{101}	1	{1}	-1	1	-1			1	-1	-1		
x_{102}	1	1	-1			1	-1	1	-1		-1	
x_{103}	1	-1	1	-1	1	-1	1	-1	1			-1
B	-3	-1	1					-1	1	1	1	1

where unfilled spaces are to be understood to hold zeros, and the pivot for the first transformation is marked by curly brackets. Further moves are as follows:

		x_{101}	x_2	x_3	x_4	x_5	x_6	x_7	x_8	x_9	x_{10}	x_{11}
x_1	1	1	-1	1	-1			1	-1	-1		
x_{102}	0	-1		-1	{1}	1	-1			1	-1	
x_{103}	2	1				-1	1			-1		-1
B	-2	1		1	-1						1	1

		x_{101}	x_2	x_3	x_{102}	x_5	x_6	x_7	x_8	x_9	x_{10}	x_{11}
x_1	1		-1		1	1	-1	1	-1		-1	
x_4	0	-1		-1	1	1	-1			1	-1	
x_{103}	2	1				-1	{1}			-1		-1
B	-2				1	1	-1				1	1

		x_{101}	x_2	x_3	x_{102}	x_5	x_{103}	x_7	x_8	x_9	x_{10}	x_{11}
x_1	3	1	-1		1		1	1	-1	-1	-1	-1
x_4	2			-1	1		1				-1	-1
x_6	2	1				-1	1			-1		-1
B		1			1		1					

It has proved possible to bring all the artificial variables to zero, and it can be inferred that a threshold element can have an assignment of weight and threshold values to make it operate as required. It is possible to derive a set of values satisfying the requirement, since the final tableau represents a solution for which all the x-variables are zero except for $x_1 = 3$, $x_4 = 2$, and $x_6 = 2$. The weight and threshold values are therefore given by:

$$\begin{aligned} w_1 &= x_1 - x_2 = 3 \\ w_2 &= x_3 - x_4 = -2 \\ w_3 &= x_5 - x_6 = -2 \\ t &= x_8 - x_7 = 0 \end{aligned}$$

If the Objective Function row had indicated (by showing non-negative entries in all positions except the first) that no further improvement was possible, before all the artificial variables had been brought to zero, it would have indicated that there was no assignment of weight and threshold values to meet the requirements. This example, though perhaps somewhat specialised in its area of application, is felt to be instructive because it shows how the first stage of the Big-M Method may be useful without the second, and also how variables not restricted to non-negative values may be introduced into Linear Programming.

REFERENCES

1. W.S. McCulloch and W. Pitts, "A logical calculus of the ideas immanent in nervous activity" **Bull. Math. Biophysics 5**, pp. 115-137 (1943)

2. N.J. Nilsson, **Learning Machines** (McGraw-Hill, New York, 1965) Ch. 5, pp. 79-94

Chapter 4

TRANSPORTATION PROBLEMS

This term is used to refer to a type of problem which can be solved by Linear-Programming methods, but for which there are better methods than the standard Simplex Method. The advantage of these methods is that they are much more economical, both in terms of computational effort and of data-storage space required. The special methods are only applicable where the problem has a particular form, so they do not supersede the Simplex Method in general.

It has been suggested by some workers that the use of the word 'Transportation' is unfortunate since it has connotations of forcible deportation of criminals to insalubrious parts of the world, and these people would prefer the appellation 'Transport Problems'. However, the word Transportation is well established by usage in this context and has been retained here.

The classical Transportation Problem is that of conveying some commodity from a number, say m, of sources to a number, say n, of destinations, with minimum cost. The amounts of the commodity available at the sources are known, as are also the amounts required at the destinations. Also, for each source-destination pair there is a known value of the cost per unit of the commodity conveyed. The solution consists of the optimal assignment of values to the amounts conveyed from each source to each destination.

The cost is calculated as a linear function of the amounts conveyed, the coefficients in the function being the unit costs already mentioned. The amounts available at sources, and amounts required at destinations, impose constraints.

A Transportation Problem may be balanced, in that the total of amounts available at sources equals that of amounts required at destinations. If the problem is unbalanced in the direction of having the total requirements greater than the total available, then no solution exists. If the imbalance is the other way round a solution can be found. It is then possible to convert the problem to a balanced one by introducing a 'dummy destination' whose requirement is the right amount to make the balance, and such that all unit costs of conveying goods to it are zero. Goods conveyed to the dummy destination are in fact left where they were. Since this trick can be used to convert an unbalanced to a balanced problem, it is only necessary to consider how to solve the balanced one.

For a balanced problem the constraints are all in the form of equations, so there is no need to introduce slack variables. For the unbalanced problem the constraints due to amounts available at the sources are impositions of upper bounds. The variables representing goods conveyed to the dummy destination are in fact the slack variables which convert these upper-bound constraints into equations.

Instead of operating on a Simplex Tableau, the standard method for Transportation Problems operates on a table having m rows corresponding to the sources and n columns corresponding to the destinations (or vice versa - the method is symmetrical with regard to sources and destinations). The main entry in a cell of the table is the amount of the commodity conveyed between the source and destination corresponding to its row and column. Also associated with it is the unit cost for that source-destination pair, which does not change. During the solution process another number is required, playing the same part as the coefficients in the Objective Function row of the Simplex tableau in determining whether a given solution is optimal and showing how it can be improved if it is not.

When a Transportation Problem is solved by hand, the best way to write it down is with the three numbers associated with a particular source-destination pair squeezed into one cell of the table. One way to do it is to write the amount of commodity transferred in the middle of the cell, and to put the other numbers in corners like the numbers in cells of a crossword puzzle. Usually, the unit cost is in one of the lower corners and the other number required is in a top corner - see Figure 4.1. The alternative is to make three separate tables, and of course this latter method tends to be favoured in books since it is easier to typeset. In the PASCAL program which will be shown, advantage is taken of the facility of PASCAL for handling records to arrange that each cell of the array storing the data in the program has the three numbers grouped together in it.

In discussing the Simplex Method it was assumed that the Objective Function was a profit to be maximised. This seemed more natural in the early stages and it was convenient to opt for one or other of maximisation and minimisation. However, it seems more natural to talk in terms of minimisation of costs when discussing Transportation Problems, so there is, here as in much of the other literature, a slightly unfortunate switch-over.

Where there are m sources and n destinations, the (m.n) entries for amounts conveyed are the variables to be optimised. Since the totals are specified for both sources and destinations, each row and each column of the table represents a constraint. It is because the constraints are superimposed on each other as rows and columns of the same table that a problem is much more concisely represented in this way than as a Simplex Tableau. The work involved in transforming the representation to correspond to successive Basic Feasible Solutions is also much less.

THE INITIAL BASIC FEASIBLE SOLUTION

An alternative way of stating the requirement for a Basic Feasible Solution is to say that the number of variables having non-zero values must not be more than the number of constraints. For the moment it will be assumed that it is equal to the number of constraints, but at a later stage it will be necessary to consider the 'degenerate' case where a

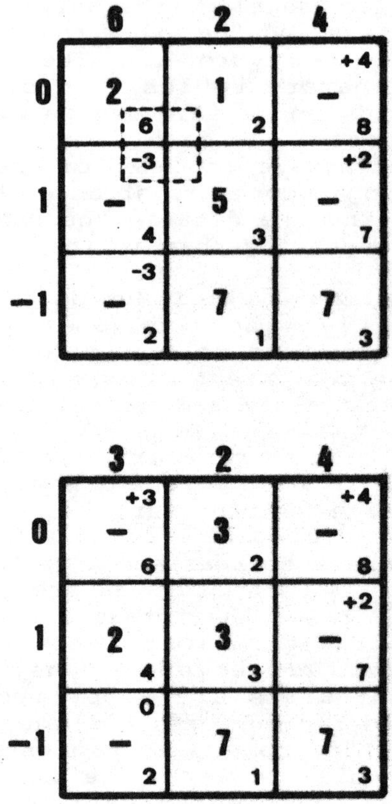

Figure 4.1. Recommended way of drawing the tables when solving Transportation Problems by hand. The large figures in the squares are amounts of goods conveyed, and those outside are row and column shadow costs. The small figures in lower right-hand corners are unit costs. Those in upper right-hand corners are differences between unit and imputed costs. The broken line shows an interchange pathway.

B.F.S. is found having fewer non-zero values. Since each row and each column represents a constraint, the number of constraints would appear to be (m+n), but in fact, for a balanced problem (to which attention is now restricted), any one constraint can be deduced if all the others are known. The number of non-zero entries at a B.F.S. is therefore restricted, not to (m+n), but to (m+n-1), the number of independent constraints.

Just as with the Simplex Method, it is necessary to find a starting B.F.S. This can be done without any regard to the unit costs, and the method which will be described and which is embodied in the sample program, is one which pays no attention to them. Such a method is simple to describe and to program, and is adequate for simple illustrative problems. An alternative which takes account of the costs is easily derived from it. The advantage of the latter is that it favours the lower-cost pathways and probably arrives at a B.F.S. having lower overall cost than that found by the simpler method. The number of subsequent steps to be taken to reach the optimum is likely to be less when the starting B.F.S. has lower cost.

The simple method is often referred to as the 'North-West Corner Rule' because it is normally applied by starting at the top left-hand corner of the table and scanning it row by row. As each cell is visited, a value is assigned to it for the amount of goods to be conveyed, the value being the largest which is possible without violating a constraint. That is to say, it is the largest which does not exceed a row or column total, having regard to amounts already assigned to other cells.

The number of non-zero entries resulting from this procedure is usually (m+n-1), because every non-zero entry completes the sum for either a row or a column (otherwise it could have been a larger entry). The final entry must complete the sum for both a row and a column, so the number of non-zero entries is normally (m+n-1). It may turn out to be less than this if an entry prior to the last one completes both a row and a column, but this possibility is being ignored for the present. The row-by-row scanning is just a simple way of visiting every cell in the table. The starting B.F.S. is likely to be closer to the optimum if, instead, cells having low unit cost are visited first, but row-by-row scanning is used here for simplicity.

The commodity to be transported may consist of indivisible objects such as ships or trucks, or of something finely divisible like coal or water or electrical power. If the amounts available and wanted have a common divisor, the quantities appearing in the optimal solution as found will be multiples of the same divisor. If there are two or more Basic Feasible Solutions which tie for first place, there will be further solutions which are linear combinations of them and which are equally good and involve quantities which are not multiples of the common divisor. It is true, however, that if the problem is presented in terms of integer numbers of, say, tonnes of coal, there will never be any advantage in dividing a tonne. The problem of what to do if the method suggests dividing a ship does not arise. Garvin [1] has suggested, frivolously, that this shows that monogamy is the best social system; there is never any advantage to be gained by dividing one's attention between two partners. He is, of course, making an unwarranted assumption of additivity of partner-satisfaction. As he is careful to point out, even if such technical quibbles as this are ignored, the argument only shows that monogamy can be the best system if partner allocation is optimal; with a different allocation of partners it can also be the worst.

Suppose the amounts of coal available at three collieries are 3, 5 and 14 tonnes, and the amounts required by three consumers are 2, 13 and 7 tonnes. This is a balanced situation, and it is possible to find a Basic Feasible Solution even before the problem has been fully stated by specifying unit costs. The table of amounts transported would be as follows, where a dash (-) is used to indicate a zero entry:

		Required		
		2	13	7
	3	2	1	-
Available	5	-	5	-
	14	-	7	7

It can be seen that the number of non-zero entries is (m+n-1), or five, as required.

MINIMISING THE COST

The unit costs are best presented as a table similar to the above. Suppose they are as follows:

6	2	8
4	3	7
2	1	3

The total cost for the B.F.S. represented can be found by multiplying corresponding entries in the two tables and summing the products.

For this B.F.S. the total cost is:

$$2 \times 6 + 1 \times 2 + 5 \times 3 + 7 \times 1 + 7 \times 3 = 57$$

Just as in the Simplex Method, the next step is to decide whether the situation can be improved by selecting one of the non-basic variables (an empty cell of the table) and letting it become basic (that is, sending some goods by that route). The first empty cell encountered in a row-by-row scan is the one at the right-hand end of the first row.

If this cell becomes occupied, it is necessary to alter the values appearing in some other cells, since the row and column totals must remain unaltered. This can be done, without affecting any other empty cell, if a closed 'interchange pathway' can be found, starting from the cell considered and returning to it, with occupied cells at all the other corners. A suitable pathway (in fact, the only one) goes from the cell considered, in position (1, 3), to (3, 3), then (3, 2), (1, 2) and back to (1, 3). This is shown by the broken line in Figure 4.1. Then, when some goods are conveyed by the route represented by (1, 3), the amounts in (3, 3) and (1, 2) must be reduced, and that in (3, 2) increased, by the same amount, in order to keep the totals as they were. The change in overall cost is equal to the amount of goods multiplied by a factor equal to the sum of the unit costs associated with (1, 3) and (3, 2), minus the sum of the unit costs associated with (3, 3) and

(1, 2). If this factor is negative, it is advantageous to let (1, 3) become occupied. The factor is an indication of the sensitivity of the overall cost to an increase in goods conveyed by the route represented by the cell, and is equivalent to an entry in the Objective Function row of a Simplex Tableau.

The factor computed from the unit costs is in fact:

$$8 + 1 - (2 + 3) = 4$$

and since it is positive there would be an increase in total cost if this cell became occupied. If it is found that the corresponding quantity is positive for all empty cells it can be inferred that no improvement is possible.

If, however, the factor is negative for one of the empty cells, the overall cost can be reduced by making changes in the entries at the corners of the interchange pathway. If the corners are numbered either clockwise or counter-clockwise, starting at the empty cell which is to become occupied, those with even numbering have their entries incremented and those with odd numbering have them reduced. The magnitude of the changes should be as large as it can be, subject to the condition that no entry in a cell can go negative. It is therefore determined by the smallest existing entry at an odd-numbered corner. The cell at that corner becomes empty when the changes are made, so the filling of one empty cell results in the emptying of at least one occupied cell.

For the next empty cell encountered in a row-by-row scan, namely that at (2, 1), the change in cost would be a multiple of the following function of the unit costs:

$$4 + 2 - (6 + 3) = -3$$

and since this is negative it is advantageous to let this cell become occupied. The odd-numbered corners of its interchange pathway are (1, 1) and (2, 2), and the smaller of the entries in these is 2. The value 2 is therefore subtracted from the entries in these two and added to the entries in (2, 1) and (1, 2). The result is:

		Required		
		2	13	7
	3	-	3	-
Available	5	2	3	-
	14	-	7	7

and it is easily verified that the row and column totals are unchanged and the overall cost has fallen to 51, from its previous value of 57.

The first step in transforming the table, then, is to decide which empty cell (representing a non-basic variable) is to become occupied (basic). For every empty cell, the factor indicating the direction of change of overall cost must be computed as above. For cells for which it is negative, an improvement can be obtained by letting the cell become an occupied one. If there are several cells for which the factor is negative, that which is most negative (largest magnitude) is chosen.

The method which was used to calculate the factor for an empty cell depended on identifying the interchange pathway with this cell at one corner and occupied cells at all the others. Provided the table has its full complement of (m+n-1) occupied cells, there is for each empty cell one, and only one, such interchange path. Having found the path, the factor was calculated as the sum of unit costs of cells at its even-numbered corners, minus the sum of unit costs for cells at odd-numbered corners. In simple examples like the 3 x 3 one used here the paths usually have four corners in all, but in larger tables they can have six, eight, or more.

It is, however, tedious to identify the interchange paths for all the empty cells, and fortunately there is a simpler way of evaluating the factors. It depends on assigning 'shadow costs' to the rows and columns of the table. These shadow costs are such that, for any occupied cell, the unit cost is the sum of the appropriate row and column shadow costs. Since there are (m+n-1) occupied cells, this is the number of conditions to be satisfied by the shadow costs, and they are (m+n) in number. One shadow cost can be assigned arbitrarily, and then the rest are evaluated by solving (m+n-1) simultaneous equations. However, the equations are such that the values can be obtained very easily. Even when the method is implemented as a computer program there is no need to invoke the standard methods for solving simultaneous equations. If we consider the original form of the table for the problem we have been considering, the shadow cost of, say, the first row can be set to the arbitrary value zero. Then, since the first two cells in this row are occupied ones, the shadow costs of the first two columns must be respectively 6 and 2 so as to sum with the row cost to the unit costs for these cells. Then, having assigned the shadow cost to column 2 it can be seen that the shadow costs for rows 2 and 3 must be 1 and -1 respectively, to let the sum be correct for the occupied cells in (2, 2) and (3, 2). Then to make it right for cell (3, 3) the shadow cost of column 3 has to be 4. The table is shown again below, with shadow costs shown instead of row and column totals, and with unit costs shown in curly brackets.

		Column Shadow Costs	
	6	2	4
0	2{6}	1{2}	-{8}
Row Shad. Costs 1	-{4}	5{3}	-{7}
-1	-{2}	7{1}	7{3}

The assignment of shadow costs was made with regard to occupied cells only. For the empty cells it is now possible to calculate a cost imputed by the shadow costs. For the empty cell at (1, 3) the imputed cost is 4, being the sum of the row and column shadow costs. This imputed cost can be seen to be the value such that, if it were the unit cost of the empty cell, there would be no change in the overall cost when values were altered around the interchange pathway. (This follows from the fact that row and column totals are unchanged, since for sets of cells whose unit costs conform to the shadow costs, total cost can be evaluated by summing row and column totals multiplied by the respective shadow costs.)

Where the unit cost for the empty cell does not equal the cost imputed to it by the shadow costs, the difference between the two is in fact the change in overall cost per unit of goods sent by the formerly-empty cell. The difference between the actual unit cost and the imputed unit cost, for an empty cell, is in fact the factor which was discussed earlier, indicating the sensitivity of the overall cost to a change in the amount of goods sent by this cell. The factors are much more easily computed on the basis of shadow costs than by identifying interchange pathways.

IDENTIFYING THE PATHWAY

As in the Simplex Method, the table represents, at any stage, a Basic Feasible Solution, and the changes correspond to transitions from B.F.S. to B.F.S. until no further improvement is possible. The first step is to work out the shadow costs and then the unit costs imputed to empty cells, and for each empty cell the difference between the actual unit cost and the imputed cost. If the differences are all positive the optimum has been reached. If one is zero and the rest non-negative an optimal solution has been reached but there is another equally good. If one or more of the cost differences is negative (actual cost less than imputed cost), improvement is possible. The cell having the most strongly negative value is chosen as the one to become an occupied cell.

Having chosen this cell, adjustments must be made to the amounts conveyed in all the cells at corners of its interchange pathway. At this stage it is necessary to identify the pathway, the task which was bypassed earlier by using the trick with row and column shadow costs. Since the whole method is to be implemented as a computer program, an algorithm for identification of the pathway is needed. There is in fact a very simple one which meets the requirement.

The empty cell which is to become occupied, along with the set of occupied cells, are the set from which the corners of the interchange pathway are to be chosen. It is possible to reduce the size of this 'candidate set' by removing from it any member which is either the only member in some row or the only one in a column, since such members can never form part of the closed pathway. These removals should be continued until no more are possible (since removing one member can mean that another comes to satisfy the condition for removal). When no more removals are possible, the members of the candidate set remaining are the corners of the required pathway. The pathway can be traced out by starting at its known member (the empty cell) and finding the other member cell in the same row, then the member cell in the same column as that cell, and so on until the empty cell is again reached.

The change in amount of goods conveyed has to be set to the smallest entry in an odd-numbered corner cell of the pathway, and it is added to the entries of even-numbered corner cells and subtracted from entries of odd-numbered ones. This results in a new form of the table, corresponding to a new Basic Feasible Solution, and the whole process is repeated until the optimum is reached. The row and column shadow costs have to be re-computed each time because the set of occupied cells is different. The method is sometimes called the 'Stepping-Stone Method' for the Transportation Problem.

COMPUTER PROGRAM

A computer program implementing the method is as follows:

```
PROGRAM transport(input, output);

LABEL 1;
TYPE  cell=  RECORD  cost, impute, goods: real;
                     candidate: boolean  END;
      which= (row, column);

VAR   rto, cto, nsource, ndest, i, j: integer;
      matrix: ARRAY[1..10,1..10] OF cell;
      rowtot, coltot, rowshad, colshad:
                               ARRAY[1..10] OF real;
      sourcetot, desttot, min: real;  demo: boolean;
      ch: char;

PROCEDURE putshadow(amount: real;
                  where, omit: integer;  how: which);
VAR i, j: integer;
BEGIN
```

```
  IF how = row THEN
  BEGIN  rowshad[where]:= amount;
    FOR j:= 1 TO ndest DO
    WITH matrix[where, j] DO
    IF (goods > 0) AND (j <> omit) THEN
      putshadow(cost-amount, j, where, column)
  END
  ELSE  BEGIN  colshad[where]:= amount;
    FOR i:= 1 TO nsource DO
    WITH matrix[i, where] DO
    IF (goods > 0) AND (i <> omit) THEN
      putshadow(cost-amount, i, where, row)
  END
END {putshadow};

PROCEDURE transform;
VAR i, j, cnt, count: integer;  min: real;

  PROCEDURE eliminate(where: integer; how: which);
  VAR i, j, cnt: integer;
  BEGIN  cnt:= 0;  IF how=row THEN
    BEGIN  FOR j:= 1 TO ndest DO
      IF matrix[where, j].candidate THEN
        cnt:= cnt + 1;
      IF cnt=1 THEN
      BEGIN  count:= count+1;  FOR j:= 1 TO ndest DO
        matrix[where, j].candidate:= false
    END  END
    ELSE  BEGIN  FOR i:=  1 TO nsource DO
      IF matrix[i, where].candidate THEN
        cnt:= cnt + 1;
      IF cnt=1 THEN
      BEGIN  count:= count+1;
        FOR i:= 1 TO nsource DO
        matrix[i,where].candidate:=  false
    END  END
  END {eliminate};

  PROCEDURE find(VAR ii, jj: integer;  how: which);
  VAR i, j, new: integer;
  BEGIN  IF how=row THEN
    BEGIN  FOR j:= 1 TO ndest DO
      IF matrix[ii, j].candidate AND (j <> jj) THEN
        new:= j;  jj:= new  END
    ELSE BEGIN  FOR i:= 1 TO nsource DO
    IF matrix[i, jj].candidate AND (i <> ii) THEN
      new:= i;  ii:= new  END
  END {find};

BEGIN  FOR i:= 1 TO nsource DO  FOR j:= 1 TO ndest DO
  WITH matrix[i,j] DO
```

```
    candidate:= (goods > 0)
      OR ((i = rto) AND (j = cto));
REPEAT  BEGIN  count:= 0;
          FOR i:= 1 TO nsource DO eliminate(i, row);
          FOR j:= 1 TO ndest DO eliminate(j, column)

        END;
UNTIL  count=0;

i:= rto;  j:= cto;  min:= -1;
REPEAT  BEGIN  find(i, j, row);
          WITH matrix[i, j] DO
          IF (min < 0) OR (goods < min) THEN
          min:= goods;
          find(i, j, column)  END;
UNTIL  (i=rto) AND (j=cto);

REPEAT  BEGIN  find(i, j, row);
          WITH matrix[i, j] DO goods:= goods-min;
          find(i, j, column);
          WITH  matrix[i,  j]  DO  goods:=  goods+min
END;
UNTIL  (i=rto) AND (j=cto);
END {transform};

PROCEDURE print;
VAR  i, j, nd: integer;
BEGIN  IF ndest > 5 THEN nd:= 5 ELSE nd:= ndest;
  writeln('Amounts of goods, and shadow costs:');
  writeln(output);
  FOR i:= 1 TO nsource DO
  BEGIN  FOR j:= 1 TO nd DO  WITH matrix[i,j] DO
   IF goods > 0 THEN write(goods:7,'  ') ELSE
   write('    -     ');
     writeln('  ',rowshad[i]:7)
  END;
  writeln(output);
  FOR j:= 1 TO nd DO write(colshad[j]:7,'  ');
  writeln(output);  writeln(output);
END {print};

PROCEDURE firstprint;
VAR  i, j, nd: integer;
BEGIN  IF ndest > 5 THEN nd:= 5 ELSE nd:= ndest;
  writeln('Cost matrix, and totals for goods:');
  writeln(output);
  FOR i:= 1 TO nsource DO
  BEGIN  FOR j:= 1 TO nd DO  WITH matrix[i,j] DO
   write(cost:7,'  ');
     writeln(' ',rowtot[i]:7)
  END;
```

```
  writeln(output);
  FOR j:= 1 TO nd DO write(coltot[j]:7,'  ');
  writeln(output);   writeln(output);
END {firstprint};

PROCEDURE northwest;
VAR  i, j, ii, jj: integer;  r, c: real;
BEGIN
  FOR i:= 1 TO nsource DO  FOR j:= 1 TO ndest DO
  BEGIN  r:= rowtot[i];
    FOR jj:= 1 TO j-1 DO  r:= r - matrix[i,jj].goods;
    c:= coltot[j];
    FOR ii:= 1 TO i-1 DO  c:= c - matrix[ii,j].goods;
    IF r < c THEN matrix[i,j].goods:= r ELSE
      matrix[i,j].goods:= c
  END
END {northwest};

BEGIN
1:  writeln(
  'Insert no. of sources (rows in matrix)');
read(nsource);
writeln(
  'Insert no. of destinations (cols in matrix)');
read(ndest);
writeln('Insert amounts available at sources');
FOR i:= 1 TO nsource DO read(rowtot[i]);
writeln('Insert amounts wanted at destinations');
FOR i:= 1 TO ndest DO read(coltot[i]);
writeln('Insert cost matrix, row by row');
FOR i:= 1 TO nsource DO  FOR j:= 1 TO ndest DO
  read(matrix[i, j].cost);
writeln('Print every step - insert y or n');
  readln(input);  read(ch);
  demo:= (ch='y') OR (ch='Y');
sourcetot:= 0;  FOR i:= 1 TO nsource DO
  sourcetot:= sourcetot + rowtot[i];
desttot:= 0;  FOR j:= 1 TO ndest DO
  desttot:= desttot + coltot[j];
IF desttot > sourcetot THEN
  BEGIN  writeln(
   'Insufficient goods at sources - start again');
    GOTO 1  END;
IF sourcetot > desttot THEN
  BEGIN  writeln(
    'Extra destination added to balance problem');
    ndest:= ndest + 1;
    coltot[ndest]:= sourcetot - desttot;
    FOR i:= 1 TO nsource DO matrix[i, ndest].cost:= 0
  END;
```

```
firstprint;
northwest;
REPEAT  BEGIN
  putshadow(0, 1, 0, row);
  rto:= -1;  min:= 0;
  FOR i:= 1 TO nsource DO  FOR j:= 1 TO ndest DO
  WITH matrix[i,j] DO
  IF goods = 0 THEN
  BEGIN  impute:= rowshad[i] + colshad[j];
    IF (rto < 0) OR (cost-impute < min) THEN
    BEGIN  rto:= i;  cto:= j;  min:= cost - impute
END
  END;
  IF demo OR (min >= 0) THEN print;
  IF min < 0 THEN transform
  END;
UNTIL  min >= 0
END.
```

It can be seen that the table representing the current B.F.S. is stored as an array of variables of type 'cell'. Each cell has four fields, of which one holds the unit cost, another the imputed unit cost, another the amount of the commodity conveyed by that route ('goods'), and a Boolean variable 'candidate' which becomes 'true' when the cell is a candidate to be included in an interchange pathway. The variable type 'which' is useful to allow procedures to be called so as to operate on either rows or columns.

PROCEDURE putshadow computes the row and column shadow costs, calling itself recursively until they have all been assigned. It is called near the end of the program with the parameters (0, 1, 0, row), which instruct it to assign the value 0 to the shadow cost for row 1. It then searches row 1 for occupied cells, which would allow it to assign some column shadow costs. It is instructed, by the assignment of value 0 to 'omit', to ignore column 0 in this search. Since there is no column so numbered, this has no effect on the search, but in the recursive calls it is necessary to give a value to 'omit' to ensure that the process does not return to the column from which a row was reached or the row from which a column was reached.

PROCEDURE transform carries out the transformation of the table, following the selection of the empty cell which has to become occupied. Its row and column are indicated by 'rto' and 'cto'. This procedure has two others declared within it -

PROCEDURE eliminate and PROCEDURE find. The first eliminates members from the candidate set of cells, incrementing the variable 'count' each time it does so. It is called repeatedly until its application to all the rows and columns leaves the value of 'count' at 0. The second, PROCEDURE find, is used in traversing interchange pathways. PROCEDURE northwest operates to find a starting Basic Feasible Solution.

The output of this program for the simple example considered is as follows:

```
Cost matrix, and totals for goods:

 6.0E+00   2.0E+00   8.0E+00    3.0E+00
 4.0E+00   3.0E+00   7.0E+00    5.0E+00
 2.0E+00   1.0E+00   3.0E+00    1.4E+01

 2.0E+00   1.3E+01   7.0E+00

Amounts of goods, and shadow costs:

 2.0E+00   1.0E+00      -        0.0E+00
    -      5.0E+00      -        1.0E+00
    -      7.0E+00   7.0E+00    -1.0E+00

 6.0E+00   2.0E+00   4.0E+00

Amounts of goods, and shadow costs:

    -      3.0E+00      -        0.0E+00
 2.0E+00   3.0E+00      -        1.0E+00
    -      7.0E+00   7.0E+00    -1.0E+00

 3.0E+00   2.0E+00   4.0E+00
```

The shadow costs are shown to the right, and below, the main table. The convention was that rows correspond to sources and columns to destinations. The result shows that an optimal solution is to send 3 units of the commodity from source 1 to destination 2, 2 units from source 2 to destination 1, and 3 units from source 2 to destination 2. From source 3, 7 units are sent to each of destinations 2 and 3.

DEGENERACY

It has been assumed that the number of occupied cells in the table is indeed (m+n-1), although it was mentioned in passing that solutions may exist for which fewer than this number are occupied. Such solutions are said to be degenerate, and they may appear either as the initial B.F.S. or after some transformations.

The method as described depends on having exactly (m+n-1) occupied cells; if there are fewer it will not always be possible to find the required interchange pathways. It is clear that degeneracy has to be avoided.

Degeneracy can be rendered impossible by slightly perturbing the problem from that presented, and solving the perturbed version. The perturbations are multiples of a small quantity, called e, which is never evaluated; it is just a vanishingly small number. The solution to the original problem is finally obtained by letting e go to zero.

Suppose there are again three sources and three destinations, but this time the amounts available are 2, 3 and 4 units, and the amounts required are 1, 1 and 7 units. Then the application of the N-W Corner Rule yields the following:

		Required		
		1	1	7
	2	1	1	-
Available	3	-	-	3
	4	-	-	4

in which there are only four occupied cells instead of the five required.

The perturbation which overcomes the difficulty consists of adding e to each 'available' amount, and adding to only the first of the 'required' amounts a sufficient multiple of e to maintain the balance, namely e multiplied by the number of sources. Alternatively, since sources and destinations are interchangeable, e can be added to each required amount, with the first available amount increased by e times the number of destinations. The table is then filled according to the N-W Corner Rule. Care must be taken in applying it, remembering that e is a vanishingly small quantity and that a negative multiple of e may form part of a combined value which is positive overall.

The table for the perturbed problem is as follows:

		Required		
		1+3e	1	7
	2+e	1+3e	1-2e	-
Available	3+e	-	2e	3-e
	4+e	-	-	4+e

The method proceeds as before but maintaining the e-values at every stage. This trick provides a complete safeguard against degeneracy. It is easy to see that it prevents its appearance in the application of the N-W Corner Rule or some other such rule. Degeneracy becomes apparent if some entry prior to the last completes both a row and a column total. This can happen if the sum of some subset (less than the total) of amounts available equals the sum of some subset of the amounts required. This cannot happen for the perturbed version since the sum of a subset of the required amounts must contain either 3e or no multiple of e at all, whereas the sum of a subset of the available amounts must contain a multiple of e intermediate between these two possibilities.

A modification of the earlier program, guarding against degeneracy, is as follows. TYPE pert is meant to convey 'perturbed total' for rows and columns:

```
PROGRAM trandeg(input, output);

LABEL 1;
TYPE  cell=  RECORD  cost, impute, goods: real;
                     egoods: integer;
                     candidate: boolean  END;
      pert=  RECORD  goods: real;
                     egoods: integer  END;
      which= (row, column);

VAR   rto, cto, nsource, ndest, i, j: integer;
      matrix: ARRAY[1..10,1..10] OF cell;
      rowtot, coltot: ARRAY[1..10] OF pert;
      rowshad, colshad: ARRAY[1..10] OF real;
      sourcetot, desttot, min: real;
      demo: boolean;  ch: char;

PROCEDURE putshadow(amount: real;
                  where, omit: integer;  how: which);
VAR i, j: integer;
BEGIN
  IF how = row THEN
  BEGIN  rowshad[where]:= amount;
    FOR j:= 1 TO ndest DO
    WITH matrix[where, j] DO
    IF ((goods > 0) OR (egoods > 0)) AND (j <> omit)
    THEN putshadow(cost-amount, j, where, column)
  END
  ELSE  BEGIN  colshad[where]:= amount;
    FOR i:= 1 TO nsource DO
    WITH matrix[i, where] DO
    IF ((goods > 0) OR (egoods > 0)) AND (i <> omit)
    THEN putshadow(cost-amount, i, where, row)
  END
END {putshadow};

PROCEDURE transform;
VAR i, j, cnt, count: integer;  min: pert;

  PROCEDURE eliminate(where: integer; how: which);
  VAR i, j, cnt: integer;
  BEGIN  cnt:= 0;  IF how=row THEN
    BEGIN  FOR j:= 1 TO ndest DO
      IF matrix[where, j].candidate THEN
        cnt:= cnt + 1;
      IF cnt=1 THEN
      BEGIN  count:= count+1;  FOR j:= 1 TO ndest DO
        matrix[where, j].candidate:= false
    END  END
    ELSE  BEGIN  FOR i:=  1 TO nsource DO
      IF matrix[i, where].candidate THEN
```

```
        cnt:= cnt + 1;
    IF cnt=1 THEN
      BEGIN  count:= count+1;
        FOR i:= 1 TO nsource DO
        matrix[i,where].candidate:=  false
    END  END
  END {eliminate};

  PROCEDURE find(VAR ii, jj: integer;  how: which);
  VAR i, j, new: integer;
  BEGIN  IF how=row THEN
    BEGIN  FOR j:= 1 TO ndest DO
      IF matrix[ii, j].candidate AND (j <> jj) THEN
        new:= j;  jj:= new  END
    ELSE BEGIN  FOR i:= 1 TO nsource DO
    IF matrix[i, jj].candidate AND (i <> ii) THEN
      new:= i;  ii:= new  END
  END {find};

BEGIN  FOR i:= 1 TO nsource DO  FOR j:= 1 TO ndest DO
  WITH matrix[i,j] DO
    candidate:= (goods > 0) OR (egoods > 0)
      OR ((i = rto) AND (j = cto));
REPEAT  BEGIN  count:= 0;
          FOR i:= 1 TO nsource DO eliminate(i, row);
          FOR j:= 1 TO ndest DO eliminate(j, column)
END;
UNTIL  count=0;

i:= rto;  j:= cto;  min.goods:= -1;
REPEAT  BEGIN  find(i, j, row);
          WITH matrix[i, j] DO
          IF (min.goods < 0) OR (goods < min.goods)
            OR ((goods = min.goods) AND
            (egoods < min.egoods)) THEN
            BEGIN  min.goods:= goods;
              min.egoods:= egoods END;
          find(i, j, column)  END;
UNTIL  (i=rto) AND (j=cto);

REPEAT  BEGIN  find(i, j, row);
          WITH matrix[i, j] DO
            BEGIN goods:= goods-min.goods;
              egoods:= egoods-min.egoods  END;
          find(i, j, column);
          WITH matrix[i, j] DO
            BEGIN goods:= goods+min.goods;
              egoods:= egoods+min.egoods  END  END;
UNTIL  (i=rto) AND (j=cto);
END {transform};
```

```
PROCEDURE print;
VAR  i, j, nd: integer;
BEGIN  IF ndest > 5 THEN nd:= 5 ELSE nd:= ndest;
  writeln('Amounts of goods, and shadow costs:');
  writeln(output);
  FOR i:= 1 TO nsource DO
  BEGIN  FOR j:= 1 TO nd DO  WITH matrix[i,j] DO
    IF (goods > 0) OR (egoods > 0) THEN
    write(' ',goods:7,' ') ELSE write('    -     ');
    writeln('  ',rowshad[i]:7);
    FOR j:= 1 TO nd DO  WITH matrix[i,j] DO
    IF (goods > 0) OR (egoods > 0) THEN
    write('      (',egoods:2,')')
    ELSE write('          ');
    writeln(output)
  END;
  writeln(output);
  FOR j:= 1 TO nd DO write(colshad[j]:7,'  ');
  writeln(output);  writeln(output);
END {print};

PROCEDURE firstprint;
VAR  i, j, nd: integer;
BEGIN  IF ndest > 5 THEN nd:= 5 ELSE nd:= ndest;
  writeln('Cost matrix, and totals for goods:');
  writeln(output);
  FOR i:= 1 TO nsource DO
  BEGIN  FOR j:= 1 TO nd DO  WITH matrix[i,j] DO
    write(cost:7,'  ');
    writeln(' ',rowtot[i].goods:7,
                   ' (',rowtot[i].egoods:2,')')
  END;
  writeln(output);
  FOR j:= 1 TO nd DO
    write(' ',coltot[j].goods:7,' ');
  writeln(output);
  FOR j:= 1 TO nd DO
    write('     (',coltot[j].egoods:2,')');
  writeln(output);  writeln(output)
END {firstprint};

PROCEDURE northwest;
VAR  i, j, ii, jj: integer;  r, c: pert;
BEGIN
  FOR i:= 1 TO nsource DO  FOR j:= 1 TO ndest DO
  BEGIN  r:= rowtot[i];
    FOR jj:= 1 TO j-1 DO
    BEGIN  r.goods:= r.goods - matrix[i,jj].goods;
      r.egoods:= r.egoods - matrix[i,jj].egoods  END;
    c:= coltot[j];
    FOR ii:= 1 TO i-1 DO
```

```
    BEGIN  c.goods:= c.goods - matrix[ii,j].goods;
      c.egoods:=c.egoods - matrix[ii,j].egoods  END;
    IF (r.goods < c.goods) OR
         ((r.goods = c.goods) AND
         (r.egoods < c.egoods))
    THEN BEGIN matrix[i,j].goods:= r.goods;
           matrix[i,j].egoods:=r.egoods  END
    ELSE BEGIN  matrix[i,j].goods:= c.goods;
           matrix[i,j].egoods:=c.egoods  END
  END
END {northwest};

BEGIN
1:  writeln(
'Insert no. of sources (rows in matrix)');
read(nsource);
writeln(
'Insert no. of destinations (cols in matrix)');
read(ndest);
writeln('Insert amounts available at sources');
FOR i:= 1 TO nsource DO read(rowtot[i].goods);
writeln('Insert amounts wanted at destinations');
FOR i:= 1 TO ndest DO read(coltot[i].goods);
writeln('Insert cost matrix, row by row');
FOR i:= 1 TO nsource DO  FOR j:= 1 TO ndest DO
  read(matrix[i, j].cost);
writeln('Print every step - insert y or n');
  readln(input);  read(ch);
  demo:= (ch='y') OR (ch='Y');
sourcetot:= 0;  FOR i:= 1 TO nsource DO
  sourcetot:= sourcetot + rowtot[i].goods;
desttot:= 0;  FOR j:= 1 TO ndest DO
  desttot:= desttot + coltot[j].goods;
IF desttot > sourcetot THEN
  BEGIN  writeln(
    'Insufficient goods at sources - start again');
    GOTO 1  END;
IF sourcetot > desttot THEN
  BEGIN writeln(
    'Extra destination added to balance problem');
    ndest:= ndest + 1;
    coltot[ndest].goods:= sourcetot - desttot;
    FOR i:= 1 TO nsource DO matrix[i, ndest].cost:= 0
  END;

FOR i:= 1 TO nsource DO rowtot[i].egoods:= 1;
coltot[1].egoods:= nsource;
FOR j:= 2 TO ndest DO coltot[j].egoods:= 0;
firstprint;
northwest;
REPEAT  BEGIN
```

```
  putshadow(0, 1, 0, row);
  rto:= -1;  min:= 0;
  FOR i:= 1 TO nsource DO  FOR j:= 1 TO ndest DO
  WITH matrix[i,j] DO
  IF (goods = 0) AND (egoods = 0) THEN
  BEGIN  impute:= rowshad[i] + colshad[j];
    IF (rto < 0) OR (cost-impute < min) THEN
    BEGIN  rto:= i;  cto:= j;
      min:= cost - impute  END
  END;
  IF demo OR (min >= 0) THEN print;
  IF min < 0 THEN transform
  END;
UNTIL  min >= 0
END.
```

A print-out from this, for the above problem exhibiting degeneracy, and with the same unit costs as used previously, is as follows. The numbers to be multiplied by e are shown in brackets:

```
Cost matrix, and totals for goods:

 6.0E+00   2.0E+00   8.0E+00    2.0E+00 ( 1)
 4.0E+00   3.0E+00   7.0E+00    3.0E+00 ( 1)
 2.0E+00   1.0E+00   3.0E+00    4.0E+00 ( 1)

  1.0E+00   1.0E+00   7.0E+00
    ( 3)      ( 0)      ( 0)

Amounts of goods, and shadow costs:

  1.0E+00   1.0E+00      -        0.0E+00
    ( 3)     (-2)
     -      0.0E+00   3.0E+00     1.0E+00
             ( 2)      (-1)
     -         -      4.0E+00    -3.0E+00
                       ( 1)
 6.0E+00   2.0E+00   6.0E+00

Amounts of goods, and shadow costs:

  1.0E+00   1.0E+00      -        0.0E+00
    ( 1)     ( 0)
  0.0E+00      -      3.0E+00    -2.0E+00
    ( 2)               (-1)
     -         -      4.0E+00    -6.0E+00
                       ( 1)
 6.0E+00   2.0E+00   9.0E+00
```

Amounts of goods, and shadow costs:

```
    -        1.0E+00   1.0E+00     0.0E+00
                ( 0)        ( 1)
  1.0E+00      -       2.0E+00    -1.0E+00
     ( 3)                   (-2)
    -          -       4.0E+00    -5.0E+00
                            ( 1)
 5.0E+00   2.0E+00   8.0E+00
```

It should perhaps be emphasised that the method can be used for much more complex problems, and that the number of sources need not be equal to the number of destinations. The program as written allows for up to ten sources and ten destinations, but larger numbers could be substituted. In most computers, the available storage space would not be exceeded unless very much larger numbers were inserted.

REFERENCE

1. W.W. Garvin, **Introduction to Linear Programming** (McGraw-Hill, New York, 1960), p. 116

Chapter 5

ASSIGNMENT PROBLEMS

Just as Transportation Problems are a special sub-class of problems solvable by Linear Programming, Assignment Problems are in turn a special sub-class of Transportation Problems. An Assignment Problem is simply a Transportation one in which the amount of the commodity available from each source, and the amount required at each destination, is one unit. (Since units are arbitrary, this is equivalent to saying the amounts are all the same.) An Assignment Problem may be balanced, with equal numbers of sources and destinations, or it may have more sources than destinations. As for the Transportation Problem, the first step is to balance the situation, if necessary, by introducing enough dummy destinations to equalise the numbers.

It is, of course, not necessary that the physical situation giving rise to a Transportation Problem has anything to do with transport; all that is necessary is that the constraint equations are in a particular form allowing two sets of them to be overlaid as rows and columns of the table. In one much-discussed study the problem was how best to arrange for table napkins from a number of functions in a catering establishment to be laundered and returned in time for a number of other functions, where a number of different laundry services were available, those which gave the faster returns being more expensive.

Assignment Problems, as the name suggests, frequently arise in contexts well removed from transport. The problem could be one of assignment of candidates to jobs, or students to laboratory benches, or football fixtures to available dates in the calendar.

Certainly, an Assignment Problem can be solved by the method described in the last Chapter, but only by employing the second version in which there are safeguards against the effects of degeneracy. It is easy to see that, if the situation was not perturbed to be non-degenerate, the application of the N-W Corner Rule would lead to a table having entries of unity in the main diagonal and zeros everywhere else, and thus only n occupied cells (for an n x n table) instead of the (2n-1) needed. Introducing perturbations proportional to e can produce the required number of occupied cells, and eventually the problem can be solved, but in following the steps it is found that very many of the transformations of the table have the effect of moving around multiples of e. They therefore do not alter the unperturbed solution. It is easy to get the feeling that there should be a more efficient method of solution, and a better method will now be introduced.

The 'Hungarian Method' can in fact be applied to the general Transportation Problem but is most readily introduced as a method for Assignment. Any method for the Assignment Problem can be extended to the general case, since there must always be some common divisor of the amounts available and required. If this is adopted as the unit, each source and destination can be split up into a set of unit sources or destinations. The problem is then one of Assignment, though usually with an enormous number of sources and destinations. There are forms of representation which allow these to be treated economically.

The Hungarian Method uses a quite different approach from that of the Simplex and Stepping-Stone Methods. It does not move from B.F.S. to B.F.S., but operates directly on the matrix of unit costs. It is convenient to continue to talk in terms of costs rather than profits, even though many of the practical applications, like assignment of candidates to jobs, are ones which are more naturally considered as producing profit. Profit can be represented as negative cost.

An important principle on which the method depends is that the optimal solution is unaffected by adding any fixed amount to any row, or to any column, of the cost matrix of a balanced Transportation Problem. Clearly the overall cost is affected, but not the amounts of goods to be sent by the different routes in order to minimise the cost. Since the row and column totals are fixed, any change in unit costs which is the same throughout a row or column must change the overall cost of any Feasible Solution by an amount equal to the total of goods for the row or column times the change in unit cost. Since the change in overall cost is the same for any Feasible Solution, the choice of the optimal solution is unaffected.

The first step in applying the Hungarian Method is to convert the cost matrix into one having only non-negative entries. This can be done by subtracting from the elements of each row an amount equal to the (algebraically) smallest value in the row. Then each row comes to hold at least one zero entry and all other entries are positive.

Rows and columns are interchangeable for this purpose, and the operation just described could equally well have been carried out on columns rather than rows. Since it is useful to have as many zero-entries as possible in the matrix, it is best to perform the operation twice, once on the rows and once on the columns. (Of course, it can equally well be done on the columns first and then the rows. The resulting matrix is also a suitable starting-point, but is not usually identical to that resulting from rows-then-columns.)

Where there are zeros in the cost matrix and all other entries are positive, an optimal solution will have been found if the only routes used to transfer goods are those with zero cost. It is therefore necessary to look for a set of 'independent' zeros. A zero is independent if there is no other zero of the set in the same row or column. (There can be other zeros in the same row or column, provided they are not included in the independent set.) If the independent set contains as many zeros as there are rows or columns, an optimal solution is indicated; the goods are conveyed via routes represented by cells of the set.

Given a sprinkling of zeros over the matrix, it

is necessary to select an independent set, preferably a maximal one. A fairly simple selection procedure can be used. It usually finds a maximal set, though this is not guaranteed. However, it can be guaranteed to produce a set which is 'complete' in the sense that no further zero element can be added to it without violating the condition of independence.

If the independent set is not large enough to indicate a solution (that is, it does not have n elements, where n is the length of side of the matrix), it is useful to be able to decide whether the set is as large as it can be or whether a better selection of zeros can be made. Use is made of the König-Egeváry Theorem (Krekó [1]), which states that the maximum number of independent zeros in a square matrix is equal to the minimum number of lines required to cover all the zeros.

The theorem has been stated in terms of zeros, but of course it applies to cells of a square matrix distinguished in any way. A 'covering line' is a line through a row or column of the matrix, and a zero is covered if one or more such lines passes through it. The value of the theorem arises from the fact that it equates two numbers which are approached from opposite directions. It refers to the minimum set of covering lines, but the maximum set of independent zeros. If the two numbers are made to agree, it can be inferred that the covering-line set is indeed minimal and the independent-zero set is maximal. The theorem will be proved in the following discussion.

Where a set of independent zeros has been found, but contains fewer than n members, the next step is to start drawing covering lines. When this is done in a particular way it will be discovered whether the selection of independent zeros can be improved. If it can, the selection is appropriately modified and if there are still less than n members the drawing of covering lines is re-commenced. If the selection is as good as it can be, the cost matrix must be modified to introduce fresh zeros. A flow diagram of the process is shown in Figure 5.1. (Although the use of flow diagrams was somewhat deprecated in Chapter 1, they are sometimes useful.)

The various components of the process will now be discussed in more detail. In the descriptions of

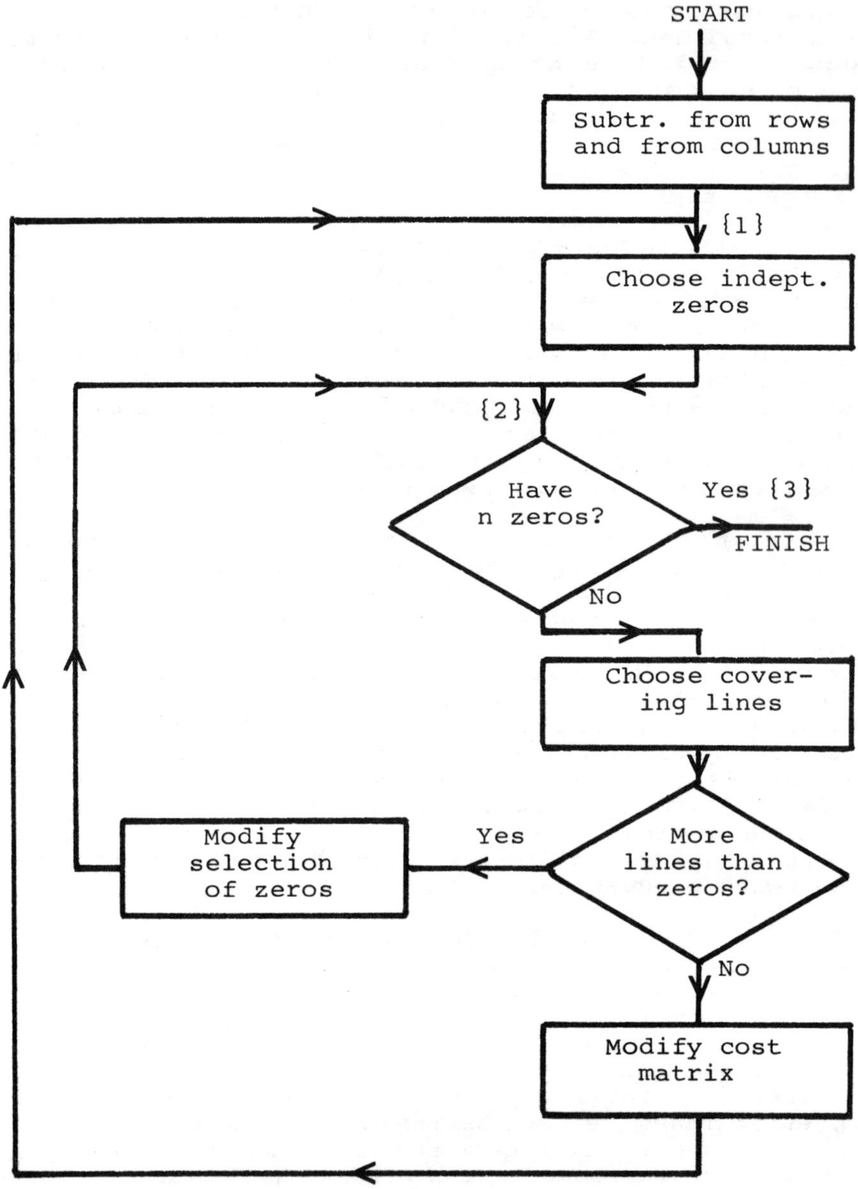

Figure 5.1. Flow diagram for the Hungarian Method. Numbers in curly brackets refer to program labels.

procedures it is convenient to have a word to mean 'row or column', and the word 'line' will be used for this, just as sociologists make much use of the word sib to mean brother or sister.

CHOOSING THE INDEPENDENT SET

The method used to choose the independent set is very simple. Every zero is a component of two lines, and if it is the only zero in one of them it should be selected if possible, otherwise the selected set cannot come to be of size n. The procedure adopted is simply to choose a line containing only one zero, or failing this as few as possible, and to let a zero in this line be introduced into the selection. Then the two lines containing this zero are, for the present purpose, deleted from the matrix, and the process is repeated as often as possible. The set of zeros may not be maximal in size but it is complete in the sense that no further zero in the matrix can be added to it without violating the condition of independence.

Suppose the original cost matrix is as follows:

5	3	4	2	6
8	3	5	5	4
2	5	3	6	8
4	2	8	3	6
3	6	9	5	3

Subtracting the smallest element, first from each row and then from each column, gives the following:

3	1	2	0	4
5	0	2	2	1
0	3	1	4	6
2	0	6	1	4
0	3	6	2	0

3	1	1	0	4
5	0	1	2	1
0	3	0	4	6
2	0	5	1	4
0	3	5	2	0

A set of independent zeros which could be chosen by the method described is shown below, marked by curly brackets. It contains four members:

3	1	1	{0}	4
5	{0}	1	2	1
{0}	3	0	4	6
2	0	5	1	4
0	3	5	2	{0}

With a small matrix like this it is not difficult to feel satisfied, from visual inspection, that this independent set is as large as is possible. It is not the only set of this size which can be found, but it is maximal. For larger matrices, or for matrices of any size when processed by a computer program, it is necessary to have a rigorous method of deciding whether a larger set is possible. This depends on the König-Egeváry Theorem and hence on covering lines.

FINDING COVERING LINES

The procedure used to draw covering lines is as follows. First, lines which contain no zero in the independent set, but do contain at least one other zero, are marked (with *). Then rows sharing a zero with a marked column, and columns sharing a zero with a marked row, are taken as covering lines. Then the marked and covering lines are deleted, for the present purpose, and the procedure repeated on the reduced matrix until there are no further lines qualifying to be marked. Sometimes it is then necessary to add a few more covering lines because of zeros of the independent set having no other zeros in the same line or column.

For the present example, the lines to be marked, and the initial covering lines, would be as follows:

```
           ¦   *
       3   1   1  {0}  4
           ¦
       5  {0}  1   2   1
           ¦
  ----{0}--3---0---4---6----
           ¦
     * 2   0   5   1   4
           ¦
       0   3   5   2  {0}
           ¦
```

Then in the reduced matrix the first column would be marked and the bottom row would become a covering line. A further covering line, either the row or the column through it, is needed to cover the independent zero near the top right-hand corner - the row has been chosen, arbitrarily. There are then four covering lines and all the zeros are covered, confirming that the set of four independent zeros is as large as is possible.

```
          *   |   *
     ----3--1--1-{0}-4----
              |
         5 {0} 1  2  1
              |
     ---{0}-3--0--4--6----
              |
       * 2  0  5  1  4
              |
     ----0--3--5--2-{0}---
              |
```

MODIFYING THE MATRIX

Since the set of independent zeros is not of the the required size (n = 5 elements) to indicate an optimal solution, it is necessary to modify the cost matrix to introduce fresh zeros. The fact that the number of covering lines equals the number of independent zeros means that no independent zero is covered twice, and the following is a way of modifying the matrix and retaining all the zeros in the independent set while introducing at least one fresh zero. It may happen that some zeros which were not in the independent set will be lost. The procedure is as follows.

Find the smallest element not covered. Let e be its value. Then:

(a) Elements not covered are reduced by e.

(b) Elements covered once remain unaltered.

(c) Elements covered twice are incremented by e.

It is not obvious that these changes are 'legal' in the sense of leaving the optimum solution unchanged. In fact they are, because they are equivalent to the following pair of 'legal' operations:

(A) Add e to all elements in columns which are covering lines.

(B) Subtract e from all elements in rows which are not covering lines.

For the present example the matrix becomes as follows, and the set of independent zeros now has five members, as shown:

3	2	1	{0}	4
4	0	{0}	1	0
{0}	4	0	4	6
1	{0}	4	0	3
0	4	5	2	{0}

Following the usual convention that rows represent sources, and columns destinations, this indicates that an optimal solution is achieved by supplying destination 4 from source 1, 3 from 2, 1 from 3, 2 from 4, and 5 from 5. There is another way in which the set of independent zeros could have been chosen, and it represents another solution equally good.

IMPROVING THE SELECTION

Provided the independent set is complete (that is, such that no other zero in the matrix can be added to it), every covering line produced by the procedure described passes through a zero of the independent set. In the first stage of the procedure, when the operation is on the unreduced matrix, a covering line not passing through a zero in the independent set would necessarily have resulted from a zero (in a marked line) which is eligible to be added to the independent set, contrary to the requirement of completeness. The covering lines produced in subsequent iterations (when the matrix has been reduced) must all pass through zeros of the independent set, since they are lines containing zeros which were not selected as marked lines in the first stage.

It follows that the number of covering lines produced by the procedure is greater than the number of zeros in the independent set if, and only if, a zero in the independent set is covered twice. Each of the covering lines through it must have been inserted because of a zero in a marked line; it is not one of the covering lines drawn right at the end of the procedure to take care of zeros of the set having no other zeros in their rows or columns.

Whenever a zero of the set is reached by a covering line drawn because of a zero in a marked line, there is the possibility of altering the membership of the independent set. The membership of the zero just reached can be cancelled, and instead the zero in the marked line is added to the set. If the marked line already contains a zero of the independent set (as it will do if it was not marked in the first stage), the membership of this zero must be cancelled, and then the zero in a marked line which initiated the covering line through it is added to the set. Eventually the tracing-back must come to a zero in one of the original marked lines, which has no zero of the independent set in it.

In the example, the fourth row became a marked line, and the covering line (covering the second column) produced by it passed through a zero of the independent set in position (2, 2). The set of independent zeros could have been modified by cancelling the membership of (2, 2) and adding to the set the zero in the marked line, namely that at (4, 2). At a later stage column 1 became a marked line and row 5 became a covering line in consequence. The membership of cell (5, 5) in the set can be cancelled, and (5, 1) can be added. However, there is another zero of the set in row 3 and its membership has to be cancelled. The zero in a marked line which gave rise to the covering line through this one is then added to the set - it is in (3, 3). In the procedure for inserting covering lines, marked lines give rise to covering lines, the covering lines, because they cause zeros in the independent set to be removed from the matrix, give rise to further marked lines, which may give rise to further covering lines, and so on. The modification of the independent set depends on working backwards along this chain of causality.

Altering the independent set as described does not alter its size; the number of zeros admitted to membership is the same as the number whose membership is cancelled. However, if a zero of the set comes to be covered twice, the chaining-back can go two ways when the membership of the doubly-covered zero has been cancelled. This leads to an increase of one in the number of zeros in the independent set.

The König-Egeváry Theorem can be deduced from this result. It follows from the definition of independence that the number of covering lines cannot be less than the number of independent zeros in the set. The above discussion shows that if the number of covering lines is greater, provided they are inserted by the procedure described, the size of the independent set can be increased.

In the program the insertion of covering lines is interrupted as soon as any zero in the independent set comes to be covered twice. The set of independent zeros can then be modified to increase its size and the drawing of covering lines is re-commenced.

It is a little difficult to illustrate the working of this part of the process, since with small matrices it is unusual to find sets of independent zeros which are complete but are not of the largest possible size. Krekó [1] discusses the following selection of independent zeros in a 7 x 7 matrix:

```
 0   1   1   0   0  {0}  0
{0}  1   1   1   1   1   0
 1   0  {0}  1   1   1   1
 0  {0}  1   1   1   1   1
 1   1   0  {0}  0   1   0
 1   1   1   1  {0}  0   1
 1   0   0   1   1   1   1
```

The independent set is complete and has 6 members. In the first stage of inserting covering lines, the last row and last column become marked lines, and covering lines are inserted as follows:

```
           |  |              *
-----0--1--1--0--0-{0}-0----
        |  |
----{0}-1--1--1--1--1--0----
        |  |
     1  0 {0} 1  1  1  1
        |  |
     0 {0} 1  1  1  1  1
        |  |
-----1--1--0-{0}-0--1--0----
        |  |
     1  1  1  1 {0} 0  1
        |  |
   * 1  0  0  1  1  1  1
        |  |
```

The fourth row can then be marked and the first column becomes a covering line. The zero in the independent set, at the start of the second row, is now doubly covered. Its membership in the set can be cancelled, and the set-modification process can work back along the covering lines and marked lines which gave rise to them. Going back along the covering line in the first column, the zero at the start of row 4 is admitted to membership. It is then necessary to cancel the membership of the selected zero already in row 4, but the zero in a marked row, at the foot of the second column, is then admitted to membership. Going along the other covering line which was involved in the double covering, namely the second row, the zero at its right-hand end is admitted to membership. The result is as follows, where zeros which have just lost their membership are shown between colons and those which have just been admitted are marked with square brackets:

```
         |  |  |        *
    ----0--1--1--0--0-{0}-0----
        |  |  |
    ---:0:-1--1--1--1--1-[0]---
        |  |  |
        1  0 {0} 1  1  1  1
        |  |  |
      *[0]:0: 1  1  1  1  1
        |  |  |
    ----1--1--0-{0}-0--1--0----
        |  |  |
        1  1  1  1 {0} 0  1
        |  |  |
      * 1 [0] 0  1  1  1  1
        |  |  |
```

It can be seen that the number of zeros in the independent set is now seven.

COMPUTER PROGRAM

A PASCAL program to apply the Hungarian Method is as follows:

```
PROGRAM assign(input, output);

LABEL 1, 2, 3;
TYPE  cell= RECORD  cost: real;
                            selected: boolean  END;
      line= (covering, marked, neither);

VAR   matrix: ARRAY[1..10,1..10] OF cell;
      rows, cols: ARRAY[1..10] OF line;
      i, j, nsource, ndest,
                         idc, jdc, indcnt: integer;
      demo: boolean;  ch: char;

PROCEDURE normalise;
VAR i, j, k: integer;  min, x: real;
BEGIN  FOR k:= 1 TO 2 DO
  BEGIN  FOR i:= 1 TO nsource DO
    BEGIN  IF k=1 THEN min:= matrix[i, 1].cost
           ELSE min:= matrix[1, i].cost;
      FOR j:= 2 TO nsource DO
      BEGIN  IF k=1 THEN x:= matrix[i, j].cost
             ELSE x:= matrix[j, i].cost;
        IF x < min THEN min:= x
      END;
      FOR j:= 1 TO nsource DO
      BEGIN  IF k=1 THEN matrix[i, j].cost:=
                       matrix[i, j].cost - min
        ELSE matrix[j, i].cost:=
                       matrix[j, i].cost - min
      END
    END {loop in i}
  END {loop in k}
END {normalise};

PROCEDURE print(str: string);
VAR i, j: integer;  ch: char;
BEGIN  writeln(str);  writeln(output);
  FOR j:= 1 TO nsource DO
  BEGIN  CASE cols[j] OF  covering: ch:= 'c';
                          marked:   ch:= '*';
                          neither:  ch:= ' '
         END {of case};
    write('    ',ch,'     ')
  END;
  writeln(output);  writeln(output);
  FOR i:= 1 TO nsource DO
    BEGIN  FOR j:= 1 TO nsource DO
      WITH matrix[i,j] DO
      IF selected THEN  write('(',cost:7,')')
                 ELSE  write(' ',cost:7,' ');
      CASE rows[i] OF  covering: ch:= 'c';
```

```
                    marked:   ch:= '*';
                    neither:  ch:= ' '
      END {of case};
      writeln('  ',ch)
    END;
  writeln(output);   writeln(output)
END {print};

PROCEDURE finalprint;
VAR i, j: integer;
BEGIN  writeln('Final assignments:');
  writeln(output);
  FOR i:= 1 TO nsource DO
    BEGIN  FOR j:= 1 TO ndest DO
      IF matrix[i,j].selected THEN  write(' X ')
                  ELSE  write(' - ');
      writeln(output)
    END
END {finalprint};

PROCEDURE clearmargins;
VAR i: integer;
BEGIN
  FOR i:= 1 TO nsource DO
  BEGIN  rows[i]:= neither;  cols[i]:= neither  END
END {clearmargins};

PROCEDURE clearselect;
VAR i, j: integer;
BEGIN  FOR i:= 1 TO nsource DO  FOR j:= 1 TO nsource
DO
  matrix[i, j].selected:= false
END {clearselect};

PROCEDURE findind;
LABEL 1;
VAR   i, j, k, cnt, first: integer;
BEGIN  clearselect;  clearmargins;  indcnt:= 0;
  1: FOR k:= 1 TO nsource DO
  BEGIN  FOR i:= 1 TO nsource DO
    IF rows[i] = neither THEN
    BEGIN  cnt:= 0;  FOR j:= nsource DOWNTO 1 DO
      IF (cols[j] = neither) AND (matrix[i,j].cost=0)
      THEN
      BEGIN  cnt:= cnt + 1;  first:= j  END;
      IF cnt = k THEN
      BEGIN rows[i]:= marked;  cols[first]:= marked;
        matrix[i, first].selected:= true;
        indcnt:= indcnt + 1;  GOTO 1  END
    END;
    FOR j:= 1 TO nsource DO
```

```
    IF cols[j] = neither THEN
    BEGIN  cnt:= 0;  FOR i:= nsource DOWNTO 1 DO
      IF (rows[i] = neither) AND (matrix[i,j].cost=0)
      THEN
      BEGIN  cnt:= cnt + 1;  first:= i  END;
      IF cnt = k THEN
      BEGIN rows[first]:= marked;  cols[j]:= marked;
        matrix[first, j].selected:= true;
        indcnt:= indcnt + 1;  GOTO 1  END
    END
  END;
  clearmargins
END {findind};

PROCEDURE cover;
LABEL 1;
VAR   i, j, k: integer;  action, zero,
                              selzero: boolean;
BEGIN  clearmargins;
  REPEAT  action:= false;
    FOR i:= 1 TO nsource DO
    IF rows[i] = neither THEN
    BEGIN  zero:=  false;  selzero:= false;
      FOR j:= 1 TO nsource DO
      IF (cols[j] = neither) AND (matrix[i,j].cost=0)
      THEN
      BEGIN  zero:= true;
        IF matrix[i,j].selected THEN selzero:= true
      END;
      IF zero AND NOT selzero THEN
      BEGIN rows[i]:= marked;  action:= true;
        FOR j:= 1 TO nsource DO
        IF (cols[j] = neither) AND
                          (matrix[i,j].cost=0) THEN
        BEGIN  cols[j]:= covering;
          FOR k:=1 TO nsource DO
          IF (rows[k] = covering) AND
                         matrix[k, j].selected THEN
          BEGIN  idc:= k;  jdc:= j;  GOTO 1  END
        END
      END
    END;
    FOR j:= 1 TO nsource DO
    IF cols[j] = neither THEN
    BEGIN  zero:=  false;  selzero:= false;
      FOR i:= 1 TO nsource DO
      IF (rows[i] = neither) AND (matrix[i,j].cost=0)
      THEN
      BEGIN  zero:= true;
        IF matrix[i,j].selected THEN selzero:= true
      END;
```

```
      IF zero AND NOT selzero THEN
      BEGIN cols[j]:= marked;  action:= true;
        FOR i:= 1 TO nsource DO
        IF (rows[i] = neither) AND
                                 (matrix[i,j].cost=0) THEN
        BEGIN  rows[i]:= covering;
          FOR k:=1 TO nsource DO
          IF (cols[k] = covering) AND
                                matrix[i, k].selected THEN
          BEGIN  idc:= i;  jdc:= k;  GOTO 1  END
        END
      END
    END;
  UNTIL  NOT action;
FOR i:= 1 TO nsource DO  FOR j:= 1 TO nsource DO
IF matrix[i,j].selected AND (rows[i] <> covering) AND
  (cols[j] <> covering) THEN  rows[i]:= covering;
1: END {cover};

PROCEDURE exchbyrow(ii, jj: integer);
LABEL 1;
VAR i, j: integer;
BEGIN  matrix[ii, jj].selected:= false;
  FOR j:= 1 TO nsource DO
  IF (cols[j]=marked) AND (matrix[ii, j].cost=0)
  AND (j <> jj) THEN
  BEGIN  matrix[ii, j].selected:= true;
    FOR i:= 1 TO nsource DO
    IF (i <> ii) AND matrix[i,j].selected THEN
      exchbyrow(i, j);
    GOTO 1
  END;
1: END {exchbyrow};

PROCEDURE exchbycol(ii, jj: integer);
LABEL 1;
VAR i, j: integer;
BEGIN  matrix[ii, jj].selected:= false;
  FOR i:= 1 TO nsource DO
  IF (rows[i]=marked) AND (matrix[i, jj].cost=0)
  AND (i <> ii) THEN
  BEGIN  matrix[i, jj].selected:= true;
    FOR j:= 1 TO nsource DO
    IF (j <> jj) AND matrix[i,j].selected THEN
    exchbycol(i, j);
    GOTO 1
  END;
1: END {exchbycol};

PROCEDURE modify;
VAR i, j: integer;  min: real;  first: boolean;
```

```
BEGIN  min:= 0;  first:= true;
  FOR i:= 1 TO nsource DO  FOR j:= 1 TO nsource DO
  IF (rows[i] <> covering) AND (cols[j] <> covering)
    AND (first OR (matrix[i, j].cost < min)) THEN
    BEGIN  first:= false;  min:= matrix[i, j].cost
    END;
  FOR i:= 1 TO nsource DO  FOR j:= 1 TO nsource DO
  IF (rows[i] <> covering) AND (cols[j] <> covering)
    THEN
    matrix[i,j].cost:= matrix[i, j].cost - min
  ELSE IF (rows[i] = covering) AND
                                  (cols[j] = covering)
    THEN matrix[i, j].cost:= matrix[i, j].cost + min
END {modify};

BEGIN  writeln('Insert no. of sources');
  read(nsource);
writeln('Insert no. of destinations');  read(ndest);
writeln('Insert cost matrix, line by line');
FOR i:= 1 TO nsource DO
BEGIN  FOR j:=1 TO ndest DO read(matrix[i,j].cost);
  FOR j:=ndest+1 TO nsource DO matrix[i,j].cost:= 0
END;
clearmargins;  clearselect;
writeln('Print each step? - insert y or n');
readln(input);  read(ch);
demo:= (ch = 'Y') OR (ch = 'y');
IF demo THEN print('Initial matrix:');
normalise;
IF demo THEN print('Normalised matrix:');
1: findind;
2: IF indcnt = nsource THEN GOTO 3;
cover;
IF idc=0 THEN BEGIN
  IF demo THEN
          print('Selected zeros & covering lines:');
  modify;
  IF demo THEN print('Modified cost matrix:');
  GOTO 1  END
ELSE BEGIN
  IF demo THEN print('With double-covered zero');
  exchbyrow(idc, jdc);  exchbycol(idc, jdc);
  indcnt:= indcnt + 1;
  IF demo THEN print('Selected zeros exchanged:');
  GOTO 2  END;
3: finalprint
END.
```

PROCEDURE normalise operates on the cost matrix to subtract an amount equal to the smallest element from the elements of each row, and then of each column. PROCEDURE findind finds the independent set of zeros. It has the rather misleading feature of altering the description of a line to 'marked' to indicate it is (temporarily) erased. PROCEDURE cover forms the set of covering lines, and if at any stage a zero of the independent set comes to be doubly covered it assigns the row and column numbers to the global variables 'idc' and 'jdc' and exits. If double covering does not occur these variables retain the value of zero assigned to them before the procedure was called.

If a doubly-covered zero of the set has been found, PROCEDURE exchbyrow and PROCEDURE exchbycol, which are both recursive, make the necessary alterations to the selected set of independent zeros. The modifications to the cost matrix to introduce fresh zeros are done by PROCEDURE modify.

A print-out from the program, using an arbitrary 4 x 4 cost matrix, is as follows:

```
Initial matrix:


   5.0E+00   3.0E+00   4.0E+00   2.0E+00
   8.0E+00   3.0E+00   5.0E+00   5.0E+00
   2.0E+00   5.0E+00   3.0E+00   6.0E+00
   4.0E+00   2.0E+00   8.0E+00   3.0E+00
```

```
Normalised matrix:


   3.0E+00   1.0E+00   1.0E+00   0.0E+00
   5.0E+00   0.0E+00   1.0E+00   2.0E+00
   0.0E+00   3.0E+00   0.0E+00   4.0E+00
   2.0E+00   0.0E+00   5.0E+00   1.0E+00
```

```
Selected zeros & covering lines:

             c         *

   3.0E+00   1.0E+00   1.0E+00 ( 0.0E+00)  c
   5.0E+00 ( 0.0E+00)  1.0E+00   2.0E+00
 ( 0.0E+00)  3.0E+00   0.0E+00   4.0E+00   c
   2.0E+00   0.0E+00   5.0E+00   1.0E+00   *
```

Modified cost matrix:

```
              c          *

   3.0E+00    2.0E+00    1.0E+00 ( 0.0E+00)   c
   4.0E+00 ( 0.0E+00)  0.0E+00    1.0E+00
( 0.0E+00)   4.0E+00    0.0E+00    4.0E+00    c
   1.0E+00    0.0E+00    4.0E+00    0.0E+00    *
```

Final assignments:

```
 -  -  -  X
 -  -  X  -
 X  -  -  -
 -  X  -  -
```

OPERATION WITHOUT OPTIMISATION

As with the Big-M variation of the Simplex Method, there is a way in which the above method can be used to check for existence of a solution to a problem, without optimising any Objective Function. Suppose four sporting events (football matches, say) have to be arranged on different days, and the number of possible days is six. Because of other commitments of the participants, event 1 is only possible on days 1 or 3, event 2 on days 1, 2, 4 or 5, event 3 on days 2, 3 or 6 and event 4 on days 1, 3 or 5. A table in which the rows correspond to days and the columns to matches is as follows, with a cross indicating that a day-match combination is possible:

```
X  X  -  X
-  X  X  -
X  -  X  X
-  X  -  -
-  X  -  X
-  -  X  -
```

Since the method depends on having a square matrix, it is necessary to introduce two dummy matches to be played on days which are, in fact, left free. These can be any days, so two columns filled with crosses must be added to the matrix. A set of independent crosses is marked in the table by curly brackets; it may be a rather improbable one to be chosen, but it has the necessary property of completeness:

```
 X  X  - {X} X  X
 - {X} X  -  X  X
 X  -  X  X {X} X
 -  X  -  -  X  X
 -  X  -  X  X {X}
 -  - {X} -  X  X
```

The independent set of crosses can be treated as the independent set of zeros was treated earlier. Applying the method for drawing covering lines, it can be seen that the fourth row and first column should be marked and covering lines can then be drawn as follows:

```
    *  |        |  |
----X--X----{X}-X--X----
       |        |  |
    - {X} X  -  X  X
       |        |  |
----X-----X--X-{X}-X----
       |        |  |
  * -  X  -  -  X  X
       |        |  |
    -  X  -  X  X {X}
       |        |  |
    -  - {X} -  X  X
       |        |  |
```

The selected cross in the third row is doubly covered, and so the selection can be modified to obtain a larger set. The new set is as follows, where square brackets mark crosses newly admitted to membership. The one which has just lost its membership is marked with colons:

```
 X  X  - {X} X  X
 - {X} X  -  X  X
[X] -  X  X :X: X
 -  X  -  - [X] X
 -  X  -  X  X {X}
 -  - {X} -  X  X
```

The selected set now has the six members needed for it to represent a solution. The selected crosses in the first four columns indicate a possible way of arranging the four events.

VARIATIONS

The general approach used here is similar to that of Krekó [1], with what are thought to be minor improvements in the method of modifying the selection of independent zeros. A number of less rigorous treatments appear in the literature, similar to this except in that they ignore the possibility of modifying the selection. One such is in Hanssmann [2]. Apart from this, numerous variations of the method appear, including one by Silver [3] presented as a working program.

REFERENCES

1. B. Krekó, **Linear Programming** (Pitman, London 1968)

2. F. Hanssmann, **Operations Research in Production and Inventory Control** (Wiley, New York 1962)

3. R. Silver "Algorithm 27: Assignment" **Comm. of the ACM 3**, no. 11, p. 603, Nov. 1960

Chapter 6

CRITICAL PATH ANALYSIS

This widely-used technique of Operations Research differs from the Linear Programming techniques in that it does not produce a clear indication of an optimal action or set of actions. Rather, at least in its elementary form, it presents information in a convenient way as an aid to planning. The planning of large projects, such as the building of a house or other structure, is a complex task. It is of course an old problem, and a variety of special charts have been devised as aids for project managers. The best-known of these is the Gantt Chart, described for example by Thierauf [1]. In recent years Critical-Path methods, especially where computers are used for the analysis, have been very widely used.

An advantage of computer analysis is that it can, with little human effort, be repeated during execution of the project. This is useful when some of the activities have not gone according to the initial plan.

A large project can be broken down into separate activities. For example, in building a house, a first stage would be to purchase some land and to obtain (in Britain) official planning permission. These would probably not be done sequentially, since the purchaser would not wish to complete the purchase until he had obtained outline planning permission, but he would not wish to apply for full permission until he had decided to purchase. Some of the architectural drawing would also be done at this stage. For the purpose of the present analysis this set of intermingling activities will be treated as one and labelled, say, 'negotiate land'.

Once the activity of 'negotiate land' has been completed, some further activities can commence, namely 'prepare detailed drawings', 'clear land' (of trees, rocks, abandoned cars, etc.), and 'arrange main services'. Then 'excavate foundations' can begin once 'prepare detailed drawings' and 'clear land' have been completed.

The whole project can be represented as a network, or arrow diagram, and Figure 6.1 shows the first part of one referring to the building of a house. The network is a directed graph, in the modern mathematical sense of the word 'graph' - see Andrew [2]. The lines of the graph correspond to activities and the nodes to events. An event is not 'reached' until all the activities leading to it have been completed, and an activity cannot be commenced until the event from which it starts has been reached. The event from which an activity starts is termed its 'tail' event and that to which it leads is its 'head' event.

The graph must have two special nodes, representing the start and finish of the project. No activities can finish at the start node, and none can start from the finish node. It is obvious that there must be no closed loops, otherwise the project can never be completed.

Each activity takes some time to complete, and for a complex project it may not be easy to deduce the overall project time. It may also be difficult to decide which activities are critical in determining overall time and thus what might be done to reduce it. If it is to be reduced by employing more workers or machines on some of the activities, it is necessary to know which ones are critical. Employing extra resources on a non-critical activity will leave the overall time unaffected and is wasteful.

If the time required for each activity is referred to as the 'length' of the corresponding line on the graph, then the activities which are critical

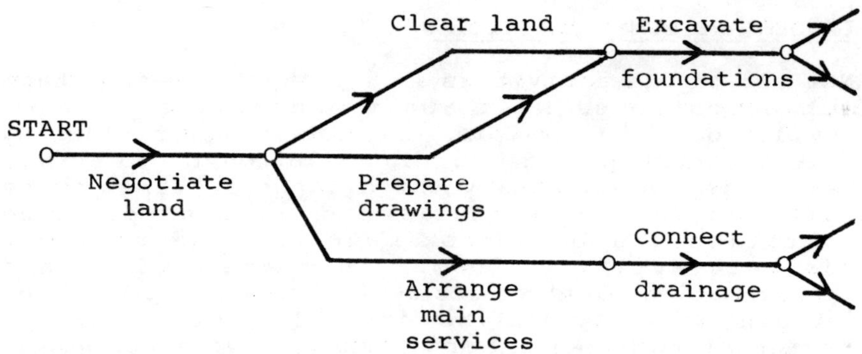

Figure 6.1. First part of activity network for building a house.

in determining overall time are those represented on the Critical Path from the start node to the finish node, and this is the path of greatest length (traversing lines in the direction of their arrows) between these nodes. A simple algorithm can be used to find the Critical Path, but for large projects the task is sufficiently complex to make the use of a computer well worth while, especially as it may be desirable to repeat the analysis at many stages during execution of the project.

A graph with six nodes is shown in Figure 6.2. The durations of activities are shown as numbers alongside the lines, and the nodes are numbered within the circles.

LADDER AND DUMMY ACTIVITIES

Not all of the activities in a project are such that subsequent ones must await a definite time of completion. For example, the obtaining of plumbing fittings must precede the installation of plumbing, but it is not necessary for all the fittings to be obtained before installation commences. The situation can be approximated in the required discrete-activity form. A rather elaborate approximation might be formed by breaking up both the 'obtain' activity and the 'install' activity into a series of components, and then to have a number of activities, like the rungs of a ladder, bridging from nodes of one series to nodes of the other. Such complication is not usually warranted, however.

A more important complication arises where the commencements of two activities must await the completion of respective sets of prior ones, and these sets are not identical but have members in common. This situation frequently arises in practice. For example, the commencement of electrical wiring must await the completion of roofing and also (we will assume, despite what was said above) completion of 'obtain electrical fittings'. Similarly the commencement of plumbing must await completion of both roofing and 'obtain plumbing fittings'.

One way to represent the situation would be by duplicating the representation of the roofing activity, as shown in Figure 6.3(a). It is best, however, to avoid multiple representation. For one

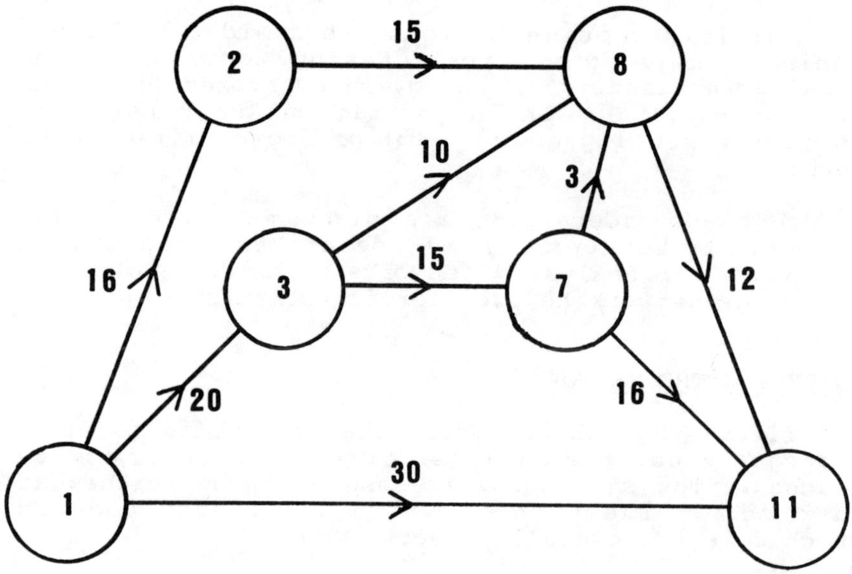

Figure 6.2. Activity network with six nodes, and activity durations marked (after Lockyer).

thing, the graph becomes unnecessarily complicated, but what is more important is that any updating of the activity time for roofing must be applied to all its representations. In more advanced forms of Critical Path Analysis, where estimates of activity time take the form of probability distributions rather than fixed values, multiple representations of an activity would be very difficult to take into account since the choices from the distribution are not independent in the multiple representations of the same activity.

Multiple representations are avoided by introducing 'dummy activities', taking zero time, and usually represented on the graph by broken lines, as in Figure 6.3(b). In large Critical Path networks there are usually quite a lot of broken lines to be seen.

These considerations are discussed in practical contexts by Lockyer [3], who describes a convenient procedure for analysing networks by hand. Figure 6.2 shows the network he uses for illustration.

ANALYSING THE NETWORK

The first stage in analysing the network is to find, for each node, the earliest time at which it can be reached. The start node is assumed to be reached at time zero. The time at which the finish node is reached is the overall project time.

The next stage is to find, for each node, the latest time at which it could be reached without extending the overall time. The difference between these earliest and latest times is termed the 'slack' for a node, and for nodes in the Critical Path it is zero. The path can therefore be identified.

It is found to be useful to evaluate three representations of the amount of extra time, or 'float' associated with an activity. These are:

(a) Total Float, the extra time which could be allowed for the activity, without extending total project time, if all other activities are completed in their stated times.

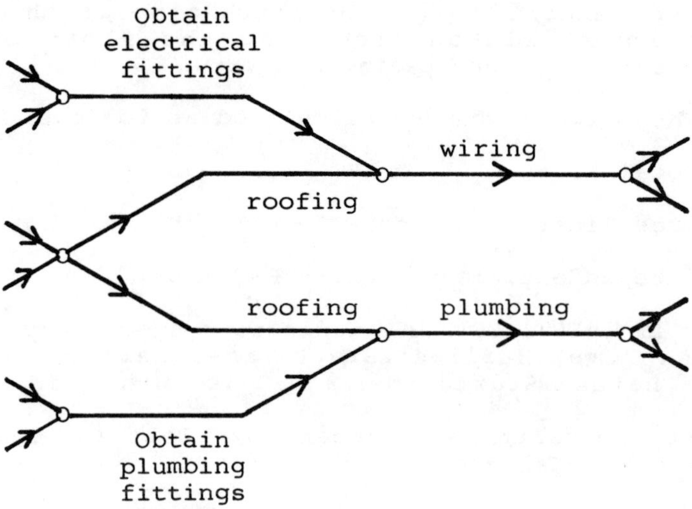

(a)

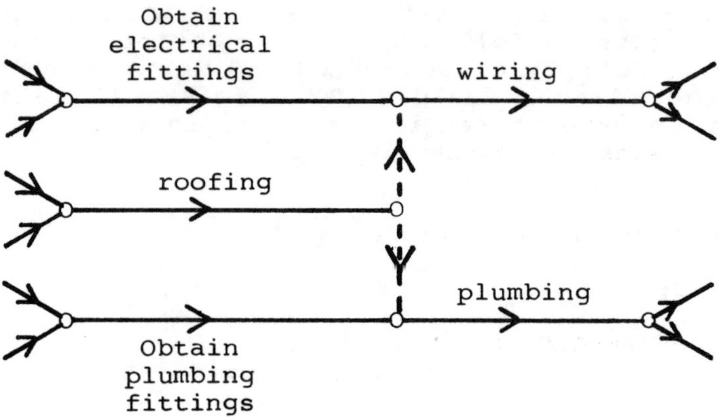

(b)

Figure 6.3. (a) Part of network with 'roofing' duplicated. (b) Equivalent without duplication, using dummy activities.

(b) Free Float, the extra time which could be allowed without reducing the float of any succeeding activity.

(c) Independent Float, the extra time which can be allowed without reducing the float of any previous or succeeding activity.

These floats can be calculated as follows:

Total Float = LFT - EST - D

Free Float = EFT - EST - D

Independent Float = EFT - LST - D

where LFT, EST, EFT, LST indicate respectively Latest Finish Time, Earliest Start Time, Earliest Finish Time, Latest Start Time, and D is Duration.

If the value for Independent Float is negative, it is taken as zero.

All the floats are zero for activities in the Critical Path.

COMPUTER PROGRAM

A computer program to do simple Critical Path Analysis is as follows. The numbering of nodes is not necessarily consecutive, and arbitrary numbers are stored as a field of TYPE node. The activity durations are stored in a matrix, in which an entry of -1 means 'no connection'.

```
PROGRAM cpa(input, output);

LABEL 2;

TYPE  node= RECORD no: integer;
                   earliest, latest: real;
                   reached: boolean  END;

VAR   activities: ARRAY[1..30,1..30] OF real;
      nodes: ARRAY[1..30] OF node;
      i, j, k, n, h, t: integer;  x, y: real;

PROCEDURE forward(from: integer);
LABEL 1;
```

```
VAR i, j: integer;  x: real;
BEGIN  nodes[from].reached:= true;
  FOR j:= 1 TO n DO
  IF activities[from, j] >= 0 THEN
  BEGIN
    x:= nodes[from].earliest + activities[from, j];
    IF nodes[j].earliest < x THEN
      nodes[j].earliest:= x;
    FOR i:=1 TO n DO
    IF (activities[i, j] >= 0) AND
                    NOT nodes[i].reached THEN GOTO 1;
    forward(j);
  1: END
END {forward};

PROCEDURE backward(from: integer);
LABEL 1;
VAR i, j: integer;  x: real;
BEGIN  nodes[from].reached:= true;
  FOR i:= 1 TO n DO
  IF activities[i, from] >= 0 THEN
  BEGIN
    x:= nodes[from].latest - activities[i, from];
    IF nodes[i].latest > x THEN
      nodes[i].latest:= x;
    FOR j:=1 TO n DO
    IF (activities[i, j] >= 0) AND
                    NOT nodes[j].reached THEN GOTO 1;
    backward(i);
  1: END
END {backward};

BEGIN  writeln('Insert no. of nodes');  read(n);
writeln('Insert node nos., start node first and');
writeln('finish node last');
FOR i:= 1 TO n DO read(nodes[i].no);
FOR i:= 1 TO n DO  FOR j:= 1 TO n DO
  activities[i, j]:= -1;
writeln('Insert activities.  For each, insert:');
writeln('Tail node no., head node no., duration');
writeln('Negative no. ends insertions');
2: read(k);
IF k > 0 THEN
BEGIN  FOR i:= 1 TO n DO
  IF nodes[i].no = k THEN t:= i;
  read(k, x);  FOR i:= 1 TO n DO
  IF nodes[i].no = k THEN h:= i;
  activities[t, h] := x;
  GOTO 2
END;
FOR i:= 1 TO n DO nodes[i].earliest:= 0;
```

```
forward(1);
FOR i:= 1 TO n DO nodes[i].latest:=
                          nodes[n].earliest;
backward(n);

writeln(output);   writeln(output);
FOR i:= 1 TO n DO
WITH nodes[i] DO
BEGIN  writeln('Node', no:3);
  writeln('Earliest', earliest:7, ' Latest',
    latest:7, ' Slack', latest-earliest:7)
END;
writeln(output);   writeln(output);
FOR i:= 1 TO n DO  FOR j:= 1 TO n DO
IF activities[i, j] >= 0 THEN
BEGIN writeln('Activity with tail',
    nodes[i].no:3, ' and head', nodes[j].no:3);
  writeln('Duration', activities[i, j]:7);
  writeln('Total float', nodes[j].latest -
    nodes[i].earliest - activities[i, j]:7);
  writeln('Free float', nodes[j].earliest -
    nodes[i].earliest - activities[i, j]:7);
  x:= nodes[j].earliest - nodes[i].latest -
    activities[i, j];
  IF x < 0 THEN x:= 0;
  writeln('Independent float', x:7);
  writeln(output)
END
END.
```

The assignment of 'earliest' times to nodes is done by the recursive operation of PROCEDURE forward. It starts from a particular node which has been reached and inspects and adjusts if appropriate the 'earliest' values of all the nodes to which this one has a connection in the forward direction. For each of these nodes, it checks whether the node has now been reached (as it has if all the nodes connecting to it have been reached), and if so the procedure calls itself recursively with this newly-reached node as parameter.

The assignment of 'latest' times is made similarly by PROCEDURE backward.

The print-out of the program on the network of Figure 6.2 is as follows:

```
Node  1
Earliest 0.0E+00 Latest 0.0E+00 Slack 0.0E+00
Node  2
Earliest 1.6E+01 Latest 2.4E+01 Slack 8.0E+00
Node  3
Earliest 2.0E+01 Latest 2.0E+01 Slack 0.0E+00
Node  7
Earliest 3.5E+01 Latest 3.5E+01 Slack 0.0E+00
Node  8
Earliest 3.8E+01 Latest 3.9E+01 Slack 1.0E+00
Node 11
Earliest 5.1E+01 Latest 5.1E+01 Slack 0.0E+00


Activity with tail  1 and head  2
Duration 1.6E+01
Total float 8.0E+00
Free float 0.0E+00
Independent float 0.0E+00

Activity with tail  1 and head  3
Duration 2.0E+01
Total float 0.0E+00
Free float 0.0E+00
Independent float 0.0E+00

Activity with tail  1 and head 11
Duration 3.0E+01
Total float 2.1E+01
Free float 2.1E+01
Independent float 2.1E+01

Activity with tail  2 and head  8
Duration 1.5E+01
Total float 8.0E+00
Free float 7.0E+00
Independent float 0.0E+00

Activity with tail  3 and head  7
Duration 1.5E+01
Total float 0.0E+00
Free float 0.0E+00
Independent float 0.0E+00

Activity with tail  3 and head  8
Duration 1.0E+01
Total float 9.0E+00
Free float 8.0E+00
Independent float 8.0E+00
```

```
Activity with tail  7 and head  8
Duration 3.0E+00
Total float 1.0E+00
Free float 0.0E+00
Independent float 0.0E+00

Activity with tail  7 and head 11
Duration 1.6E+01
Total float 0.0E+00
Free float 0.0E+00
Independent float 0.0E+00

Activity with tail  8 and head 11
Duration 1.2E+01
Total float 1.0E+00
Free float 1.0E+00
Independent float 0.0E+00
```

REFINEMENTS

Usually it is impossible to know in advance how long an activity will take, and what is perhaps the best-known implementation of Critical Path Analysis allows the durations to be expressed as probability distributions. This is the PERT algorithm, described by Robinson [4]. There appears to be some doubt about the significance of the acronym; in the early days of the development of the method it seems to have been 'Program Evaluation Research Task', but later, presumably when the method could be regarded as an established technique, the meaning seems to have shifted to 'Program Evaluation and Review Technique'. Its use is said to have produced impressive savings of time in military development projects in the U.S.A.

As was mentioned earlier, examination of a project in terms of its Critical Path is useful in connection with the allocation of resources (men, machines) to the different activities. A big contractor is likely to have several projects in hand at the same time, and is interested in allocating resources over all of them. A program allowing multi-project allocation is known as RAMPS and is described by Lambourn [5]. It is of course important to remember that the relationship between resources and completion time may be difficult to specify. If five men dig a trench in six days, it does not necessarily follow that ten would dig it in three. There may not be room for more than five men to work in the trench, so the others would either stand idle or get in the way.

Algorithms for Critical Path Analysis are fully discussed by Battersby [6] and by Carruthers and Battersby [7].

REFERENCES

1. R.J. Thierauf, **An Introductory Approach to Operations Research** (Wiley, Santa Barbara, 1978) pp. 153-155

2. A.M. Andrew, **Artificial Intelligence** (Abacus, Tunbridge Wells, 1983), pp. 58-60

3. K.G. Lockyer, **An Introduction to Critical Path Analysis** (Pitman, London, 1964)

4. F.D. Robinson "The background of the PERT algorithm" **Computer Journal 5**, pp. 297-300, 1963

5. S. Lambourn "Resource allocation and multi-project scheduling (RAMPS) - a new tool in planning and control" **Computer Journal 5**, pp. 300-304, 1963

6. A. Battersby, **Network Analysis for Planning and Scheduling** (Macmillan, London, 1964)

7. J.A. Carruthers and A. Battersby "Advances in Critical Path Methods" **Operational Research Quarterly 17**, pp. 359-380, 1966

Chapter 7

DYNAMIC PROGRAMMING

This term was introduced by Bellman [1] to denote a range of techniques applicable to the optimisation of multi-stage decision processes. The term does not lend itself to precise definition as did Linear Programming; it denotes a particular approach to problem-solving rather than a set of methods for solving problems having well-defined characteristics.

Bellman gives his 'Principle of Optimality' as follows:

> "An optimal policy has the property that, whatever the initial state and initial decision are, the remaining decisions must constitute an optimal policy with regard to the state resulting from the first decision."

The principle may at first seem to be too obvious to be worth stating. Its value lies in the fact that it indicates how decisions about the control of multi-stage processes can often be made serially. Optimisation over the whole series of decisions, regarded as a one-step parallel process, is likely to be formidable because of its high dimensionality. The Principle of Optimality often suggests methods of solution such that the computational complexity does not increase steeply with an increase in the number of decision stages.

As the time interval between decisions is made shorter, a multi-decision process comes increasingly to resemble the continuous-time control situation which is the primary concern of Control Engineering. The ideas of Dynamic Programming can therefore be applied to continuous control, but in applications in Operations Research the decisions are normally discrete and individually discernable. The general idea is applicable both in situations where the outcome of a decision is predictable (the deterministic situation) and where it depends on a chance effect whose probability distribution is known (the stochastic situation). The stochastic situation is more interesting and gives a better feeling for the value of the method. A type of problem which illustrates the approach is that of Purchasing under Uncertainty (see Thierauf [2]).

PURCHASING UNDER UNCERTAINTY

Suppose it has been decided to hold a party on a certain date, and for convenience the supply of drinks will all be purchased on the same day. Suppose there are four days on which the party-giver will be free to go shopping. On any one day he may find any one of the following:

(a) He has to pay full retail prices and his bill comes to £30.

(b) The drinks are on special offer and he can get what he wants for £25.

(c) There is an extra-special promotional offer which allows him to pay only £20.

Clearly, if he finds on the first of the four days that situation (c) holds, he should buy the drinks there and then. In fact, on any of the four days he should make the purchase if he encounters situation (c) and he has not already made the purchase. It is also clear that he should not make the purchase if he encounters situation (a) on any but the last day. However, it is not obvious what he should do at times when he has not yet made the purchase and he encounters situation (b).

It is usually assumed, in this kind of discussion, that the aim is to minimise the 'expected cost' of the purchase (or, in other contexts, to maximise an 'expected return'). The expected cost is the sum of all the possible costs, each multiplied by the probability of it becoming the actual cost. The use of 'expected' values in optimisation is discussed in the next section. It will be shown that an action which maximises the expected return may not always be that which is most desirable to a human decision-maker. However, for mathematical convenience it is usual to analyse the situation as though it were.

It is necessary to assume that the party-giver can assign probabilities of occurrence to the situations (a), (b) and (c). The probability values need not be the same for each of the four shopping days, but for simplicity it will be assumed that they are, and that situation (a) has probability 0.6, and (b) and (c) have 0.25 and 0.15 respectively.

The expected cost of making the purchase on the last day (or on any specified day), is given by (in pounds):

$$0.6 \times 30 + 0.25 \times 25 + 0.15 \times 20 = 27.25$$

In making his decision on the second-last day, therefore, the party-giver should purchase if he has not already done so and the goods are offered at less than £27.25. In other words, he should buy if he can get them at either £25 or £20, and his expected cost for the purchase, given that he has reached the second-last of the shopping days, is:

$$0.6 \times 27.25 + 0.25 \times 25 + 0.15 \times 20 = 25.6$$

This indicates that in the third-last shopping day he should only purchase if he has not already done so and the goods are offered at less than £25.60. His expected cost for a purchase in the last three days is therefore:

$$0.6 \times 25.6 + 0.25 \times 25 + 0.15 \times 20 = 24.61$$

On the first shopping day he should only purchase if the goods can be obtained for less than £24.61. In other words, he buys only if they are offered for

£20 and his expected cost is:

$$0.85 \times 24.61 + 0.15 \times 20 = 23.92$$

It can be seen, then, that the optimal policy is to purchase on the first day only if the goods can be obtained for £20, but on the next two days he will purchase, if he has not already done so, for either £20 or £25. If he has held back until the last day he must purchase whatever the price. Following this optimal policy his expected cost is £23.92.

Even this simple example would not be so easy to solve if the purchasing policy were considered as a whole instead of in four steps. The purchasing process was treated as a one-stage decision process and then the result of this was built into the solution for a two-stage process and so on. The method can easily be extended to deal with many more stages and with a choice from a greater number of alternatives at each. The alternatives need not be discrete; the method could equally well be applied if the cost-information took the form of a continuous probability distribution for each shopping day.

THE ST. PETERSBURG PARADOX

Questions relating to probability have always been of interest to gamblers. The question whether people act so as to maximise their expected return was discussed in the 18th century in terms of the 'St. Petersburg Paradox' (see, for example, Pedoe [3]). The Paradox gained its name because it was discussed at length by one of the Bernoullis in the Transactions of the St. Petersburg Academy. It is necessary to suppose that people have the opportunity to take part in a rather curious game, in which a coin is tossed as often as necessary until heads appears. If heads appears on the first toss the player is paid one pound. If it appears for the first time on the second toss he is paid two pounds, four pounds if it is on the third toss, then eight, sixteen, thirty-two and so on.

The expected return for a player partaking in this game can be shown to be infinite. The probability that heads appears on the first toss is 1/2, on the second toss it is 1/4, then 1/8, 1/16, and so on. The expected return is therefore the sum to an infinite number of terms of the series:

$$(1/2).1 + (1/4).2 + (1/8).4 + (1/16).8 + \ldots..$$

This is the sum of an infinite number of terms each equal to 1/2. Since the expected return is infinite, people who want to maximise their expected return should be willing to pay any finite amount of money to participate in this game. In fact, people do not seem to be interested in the idea of paying any large sum in order to play, and it can be argued that they do not operate so as to maximise their expected return.

The Paradox is not really very convincing. The mathematical argument showing that the expected return is infinite depends on the tacit assumption that the winnings will be paid from a limitless bank. It is rather likely that people have private doubts about the existence of such a bank. If an upper limit is placed on the amount which can be paid out, the expected return for participating becomes finite and the argument collapses.

Suppose, for example, the bank contains £1000. The sum due will be paid in full if the first appearance of heads is in the first ten throws. This follows from the fact that the sum is equal to two raised to the power (n-1) where n is the number of the final throw, and two raised to the power 9 is 512. The probability that the first appearance of heads will not be in the first ten throws is:

$$1/2^{10}, \text{ or } 1/1024$$

and so the expected return is equal to the sum of the first ten terms of the series considered earlier, plus a term equal to

$$1000(1/1024), \text{ or } 1000/1024$$

corresponding to the possibility that the first head appears later, in which case the payout is £1000. This final term contributes just under one pound to the summation for expected return. The possibility of heads in each of the first ten throws requires a contribution of 1/2 for each throw, so the total expected return is just under six pounds.

The St. Petersburg Paradox is somewhat tortuous. What is probably a better argument to show that people do not operate to maximise their expected returns is the observation that both gambling establishments and insurance companies continue to thrive. Dealing with either of these types of enterprise reduces an individual's expected financial return from the business of living. Neither book-makers nor insurance companies are philanthropists. People take out insurance in order to reduce the variability of the possible returns, because it is arranged that certain kinds of disadvantageous events will be ameliorated by a payment from the insurance company. On the other hand, gambling establishments have the opposite effect of increasing the variability of possible outcomes and such uncertainty is sometimes felt to be desirable. (Of course, some patrons of gambling establishments believe they have unbeatable systems and that the increased variability does not apply to them. There are certainly others who have no such illusions but enjoy having a 'flutter' nevertheless.)

However, these arguments are somewhat academic in connection with the problems to which Dynamic Programming is likely to be applied. The variability of possible returns is not usually vastly different over the range of solutions from which the optimum is selected. Under these conditions it is reasonable to make the choice that maximises the expected return.

DETERMINISTIC PROBLEMS

Although the most interesting applications of Dynamic Programming are to stochastic situations, it is also useful in connection with deterministic ones.

Simpson [4] discusses a number of such applications, including an interesting one having a bearing on investment policy. He relates it as the problem faced by a farmer who has a stock of wheat and who intends to farm for a certain number of years and then to sell up. In each year he can sell some of the wheat and sow the remainder, and he wants to make as much profit as possible. There is no restriction on the amount of land he can use and hence on the amount he can sow in any year. If all selling is at a fixed rate his best policy is to sell nothing until the very end, but Simpson considers the situation in which his sales in any one year go some way towards saturating the market, so that his remuneration is a convex-upward function of the amount sold. Under these conditions the problem appears quite complex, but can be solved fairly easily by a dynamic-programming approach.

Like the problem of purchasing under uncertainty, the farmer's is one in which decisions have to be made sequentially in time. In some other problem areas it is useful to let decisions be made sequentially even though the situation is not inherently sequential. An example of such a problem is that of allocating funds to a number of different areas, or projects, where the information on returns does not fit any simple mathematical model. Carlson [5] considers the allocation of advertising money in three areas in which a company operates.

Suppose a company operates in England, Scotland and Wales, and it is estimated that the returns from spending different amounts on advertising in these areas are as tabulated below. The expenditure might be in units of thousands of pounds and the return in units of millions of pounds.

England

Expenditure	0	1	2	3	4	5	6
Return	0	2	4	6	7	7	6

Scotland

Expenditure	0	1	2	3	4	5	6
Return	0	2	3	4	4	4	5

Wales

Expenditure	0	1	2	3	4	5	6
Return	0	1	2	3	4	5	5

The question is how best to spend 6 units of money on advertising.

If it assumed that the expenditure is always in integer multiples of the basic unit, there are only 14 ways of making the allocation, so it would not be very difficult to evaluate the profitability of all of them. In practice the situation is likely to be more complex because of a greater number of areas, and of alternative amounts for each allocation. It is then necessary to look for a less tedious way of finding the optimum.

Using the general idea of Dynamic Programming it is possible to consider first a one-stage process in which the decision is made for only one area, then a two-stage process, and so on. For the one-stage process in which only England is considered, the returns can be seen by inspection of the table. It would be nonsense to spend 6 units on advertising in England, since a greater return is obtained by spending less.

To consider the two-stage process in which money is allocated to both England and Scotland, a table of combined returns can be drawn:

		England						
		0	1	2	3	4	5	6
	0	0	2	4	6	7	7	6
	1	2	4	6	8	9	9	8
	2	3	5	7	9	10	10	9
Scotland	3	4	6	8	10	11	11	10
	4	4	6	8	10	11	11	10
	5	4	6	8	10	11	11	10
	6	5	7	9	11	12	12	11

This can be used to determine the maximum profit from spending 0, 1, 2, 3 .... units in the two countries. To find the maximum from spending 4 units, for example, it is simply necessary to find the biggest entry along the sloping line joining the (0, 4) cell to the (4, 0) one. The line is marked on the table, and it can be seen that the biggest entry along it has value 8.

Inspecting of other diagonal lines indicates the

maximum returns from spending other amounts of money on England-and-Scotland. They can be tabulated as follows:

England-and-Scotland

Expenditure	0	1	2	3	4	5	6
Return	0	2	4	6	8	9	10

A combined table for expenditure in all three countries can then be formed, as:

		England-and-Scotland						
		0	1	2	3	4	5	6
	0	0	2	4	6	8	9	10
	1	1	3	5	7	9	10	11
	2	2	4	6	8	10	11	12
Wales	3	3	5	7	9	11	12	13
	4	4	6	8	10	12	13	14
	5	5	7	9	11	13	14	15
	6	5	7	9	11	13	14	15

In this table, the returns for allocating 6 units in various ways lie along the sloping line marked on the table. (The lower right-hand triangle need not have been filled in if expenditure is limited to 6 units). The maximum entry along this line is 10, and it appears in three of the cells, showing there are three equally-good optimal solutions. The amount allocated to England-and-Scotland can be 4, 5 or 6 units, with the remainder going to Wales. To see how the 4, 5 or 6 units should be divided between England and Scotland it is necessary to return to the earlier square table, which shows, for example, that the best way to allocate 6 units is either to spend 3 in each country or, equally profitably, to spend 4 in England and 2 in Scotland.

This is a fairly trivial example, but it illustrates the advantage of tackling a seemingly-complex optimisation problem in a step-by-step fashion. A similar method can be used to force linear-programming problems into the dynamic-programming framework, but it is not clear that there is any advantage in so doing. In the next section another kind of stochastic situation will be considered.

THE TAXICAB PROBLEM

In most applications of Dynamic Programming it is assumed that the multi-stage decision process to be optimised has a finite and known number of stages. Dreyfus [6] has shown how to find the steady-state solution which applies in the middle of a long series of decisions. He illustrates the method by what he calls the 'Taxicab Problem'. This is the problem of finding the best policy to be followed by a taxi driver during his working day.

The system to be optimised may exist in any of a number of states. For the taxi-driver these are three towns between which he plies. In each town he may take one of three alternative actions in order to obtain his next fare. These actions are:

(1) Cruise and hope to be hailed.

(2) Drive to a taxi rank.

(3) Pull in and wait for a radio call.

For each combination of current state and action taken, there is a set of three probability values, being the probabilities of transitions to each of the three states in the next journey. These are denoted by p_{ij}, where the transition is from state i to state j, or by $p_{ij}(q)$ for the transition between these states following action q. There is also a set of expected returns, r_{ij}, associated with the possible state-transitions. The problem is to find a policy which maximises the average return per journey. This is a rule for deciding which alternative to choose in each state of the system.

The problem situation is specified by tabulating all the probability values and expected returns, as follows:

State of System (Town)	Action	Transn. probty. p_{ij}			Return r_{ij}		
		j=1	2	3	j=1	2	3
1	1	1/2	1/4	1/4	10	4	8
	2	1/16	3/4	3/16	8	2	4
	3	1/4	1/8	5/8	4	6	4
2	1	1/2	0	1/2	14	0	18
	2	1/16	7/8	1/16	8	16	8
	3	1	0	0	0	0	0
3	1	1/4	1/4	1/2	10	2	8
	2	1/8	3/4	1/8	6	4	2
	3	3/4	1/16	3/16	4	0	8

In the original presentation, town 2 was said to have no radio-call system, so action 3 was not available there. For solving the problem in a computer program it is convenient to have all possible actions available from all states of the system, so action 3 has been included with an arbitrary assignment of transition probabilities and an unfavourable assignment of returns. The latter makes it unlikely that this action will be included in the optimal policy. (In fact it is not.)

The Principle of Optimality can be stated formally as follows, where $f_n(i)$ is the optimal expected return from an n-stage process starting from state i, and N is the number of states:

$$f_n(i) = \max_q \sum_{j=1}^{N} p_{ij}(q)\,[r_{ij}(q) + f_{n-1}(j)]$$

For any fixed policy the action q to be taken in any state is pre-determined, so we can write:

$$V_i^n = \sum_{j=1}^{N} p_{ij}\,(r_{ij} + V_j^{n-1})$$

where p_{ij} and r_{ij} appear without arguments, and V_i^n is the expected return from an n-stage process starting in state i.

For simplicity, the next step of the argument will not be fully justified here, but it is hoped that it is made to look plausible. Provided any state of the system (town) can be reached from any other, the expected return for large n, starting from state i, can be expressed as:

$$V_i^n = v_i + n.g$$

where g is an average return per journey, and v_i is a contribution (positive or negative) associated with the start from state i. The condition that any state can be reached from any other ensures that the states are not divided into two or more distinct subsets with no communication between them. In such a case the above relationship would not hold as the average return per journey would depend on which subset had been entered.

Combining the last two equations and eliminating n, we have:

$$g + v_i = \sum_{j=1}^{N} p_{ij} (r_{ij} + v_j)$$

and, for a given set of values of the p_{ij} and r_{ij} this lets us write down a set of N simultaneous equations, corresponding to i = 1, 2, ... N.

In the N equations there are (n+1) unknowns, namely g and the N values v_1, v_2 .... It is permissible to set one of the v_i arbitrarily to zero and then to solve the equations for the other N unknowns. For any fixed policy, the values of g and the v_i can be derived. Dreyfus calls this part of the process the 'Value-Determination Operation'.

The process proceeds by examining some particular policy and deciding whether it is optimal. The method of testing is such that, if the optimum has not been reached a better policy is indicated. Dreyfus shows that the process must eventually find the optimum. The policy initially considered could be chosen arbitrarily. It is likely to be nearer to the optimum if it is the one which maximises the immediate returns, that is, the values obtained by summing:

$$\sum_{j=1}^{N} p_{ij} \cdot r_{ij}$$

for a state- and action-combination.

For the example considered, the policy which maximises the immediate returns is [1, 1, 1] - that is, 'always cruise'.

Substituting the appropriate values for this policy, the equations to be solved may be written as:

$$g + v_1 = 8 + 1/2\ v_1 + 1/4\ v_2 + 1/4\ v_3$$
$$g + v_2 = 16 + 1/2\ v_1 + 1/2\ v_3$$
$$g + v_3 = 7 + 1/4\ v_1 + 1/4\ v_2 + 1/2\ v_3$$

Setting v_3 to zero and rearranging, these become:

$$(1-1/2)\ v_1 - 1/4\ v_2 + g = 8$$
$$-1/2\ v_1 + v_2 + g = 16$$
$$-1/4\ v_1 - 1/4\ v_2 + g = 7$$

and solving gives $v_1 = 1.33$, $v_2 = 7.47$, $v_3 = 0$, and $g = 9.2$.

The next step is to decide whether this policy is optimal. The procedure used to make the decision, and to produce a better policy if possible, is called by Dreyfus the 'Policy-Improvement Routine'. It uses the numerical values calculated above.

If the policy is indeed optimal, the expected return from the ensuing sequence of journeys will be reduced if one decision is made contrary to the policy, even if all subsequent decisions are made according to it. A table is therefore prepared in which expected returns are shown for each combination of a state (i.e town), and an action. This is the return when this action is taken as the first on leaving the state. Subsequent decisions are taken according to the policy which has just been evaluated.

Following the first stage of the problem as stated, the table is as follows (taken from the print-out of the computer program):

State	1	altrntv.	1	1.053E+01
State	1	altrntv.	2	8.433E+00
State	1	altrntv.	3	5.517E+00
State	2	altrntv.	1	1.667E+01
State	2	altrntv.	2	2.162E+01
State	2	altrntv.	3	1.333E+00
State	3	altrntv.	1	9.200E+00
State	3	altrntv.	2	9.767E+00
State	3	altrntv.	3	5.967E+00

The word 'altrntv.' is used to denote the alternative action chosen. The numbers in the rightmost column are not actually the total return from all subsequent journeys of the taxi, but the expected return from n journeys, where n is large, minus (n-1)g. This value is independent of n, and is calculated as:

$$\sum_{j=1}^{N} p_{ij} (r_{ij} + v_j)$$

In words, this is the sum of the immediate return on starting from state i, and the sum of the contributions from starting in each of the possible states multiplied by the probability of moving to it from the starting state.

For the three returns calculated in this way for i = 1, that corresponding to action 1 is the largest. For i = 2 and i = 3, the returns corresponding to action 2 are the largest. This indicates that the next policy to be tried is [1, 2, 2]. When the new policy indicated in this way is the same as the previous one, the optimum has been reached.

In this example, the optimum policy is found to be [2, 2, 2], and as Dreyfus points out, this is actually the policy which minimises the immediate returns. Presumably he has carefully chosen the numerical values to bring this about, but nevertheless it illustrates the point that a choice made on the basis of immediate returns is often far from optimal. As he puts it, in problems of this sort the birds in the bush are often worth more than the birds in the hand.

COMPUTER PROGRAM

In order to write a computer program to carry this out, it is necessary to have a procedure for solving sets of simultaneous equations. PROCEDURE solve in the following is a simple one for this purpose, and its working will not be explained here. The coefficients of the equations are assembled into an array called 'eqns'. The solution values are left in the rightmost column of the part of the array which is used, that is, the column which previously held the r.h.s values for the equations. The program is as follows:

```
PROGRAM taxicab(input, output);

LABEL 1;
TYPE eqarr = ARRAY[1..10, 1..11] OF real;
VAR  n, na, nn, i, j, k: integer;
       {n is no. of states, na no. of alt. actions
                  nn is no. of coefficients printed}
     profit, max, x: real;
     py, npy: ARRAY[1..10] OF integer;
       {py is policy, npy is new policy}
     p, r: ARRAY[1..10, 1..10, 1..10] OF real;
       {p is array of transition probabilities,
                                       r of returns}
     q: ARRAY[1..10, 1..10] OF real;
       {q is array of expected returns}
     eqns: eqarr;

PROCEDURE solve(VAR eq: eqarr; n: integer);
VAR i, j, k: integer;  x: real;
BEGIN  FOR i:= 1 TO n DO
  BEGIN  k:= i;  FOR j:= i+1 TO n DO
    IF abs(eq[j, i]) > abs(eq[k, i]) THEN k:= j;
    FOR j:= 1 TO n+1 DO
    BEGIN  x:= eq[i, j];  eq[i, j]:= eq[k, j];
      eq[k, j]:= x  END;
    FOR j:= i+1 TO n DO
    BEGIN  x:= eq[j, i]/eq[i, i];  eq[j, i]:= 0;
      FOR k:= i+1 TO n+1 DO
      eq[j, k]:= eq[j, k] - x*eq[i, k]  END
  END;
  FOR i:= n DOWNTO 1 DO
  BEGIN eq[i, n+1]:= eq[i, n+1]/eq[i, i];
```

```
    FOR j:= 1 TO i-1 DO
    eq[j, n+1]:= eq[j, n+1] - eq[j, i]*eq[i, n+1]
  END
END {solve};

BEGIN  writeln('Insert no. of states');  read(n);
writeln('Insert no. of alt. actions');  read(na);
nn:= 4;  IF n < nn THEN nn:= n;
FOR i:= 1 TO n DO  FOR j:= 1 TO na DO
BEGIN  writeln('State', i:2, ' alternative', j:2);
  py[i]:= 1;
  writeln('Insert', n:2, ' transn. probabilities');
  FOR k:= 1 TO n DO read(p[i, j, k]);
  writeln('Insert', n:2, ' returns');
  FOR k:= 1 TO n DO read(r[i, j, k]);
  q[i, j]:= 0;
  FOR k:= 1 TO n DO
  q[i, j]:= q[i, j] + p[i, j, k]*r[i, j, k];
  IF q[i, j] > q[i, py[i]] THEN py[i]:= j
END;

1: writeln(output);  writeln(output);
writeln('Value-determination opn.  for policy:');
FOR i:= 1 TO n DO write(py[i]:4);
writeln(output);  writeln(output);
FOR i:= 1 TO n DO  FOR j:= 1 TO n+1 DO
IF j=n THEN eqns[i, j]:= 1
ELSE IF j=n+1 THEN eqns[i, j]:= q[i, py[i]]
ELSE IF j=i THEN eqns[i, j]:= 1 - p[i, py[i], j]
ELSE eqns[i, j]:= -p[i, py[i], j];

writeln('Coeffts. of eqns. to be solved:');
FOR i:= 1 TO n DO
BEGIN  FOR j:= 1 TO nn DO write(eqns[i, j]:10,'  ');
  write('  ', eqns[i, n+1]:10);  writeln(output);
END;  writeln(output);
solve(eqns,  n);
writeln('Values of v are as follows:');
FOR i:= 1 TO n-1 DO writeln(eqns[i, n+1]:10);
writeln(0:10);
writeln('Average profit/trip', eqns[n, n+1]:10);
writeln(output);  writeln(output);

writeln('Policy-Improvement Routine');
writeln(output);
FOR i:= 1 TO n DO
BEGIN  max:= 0;  npy[i]:= 1;
  FOR j:= 1 TO na DO
  BEGIN  write('State', i:3, ' altrntv.', j:3,
                                           '  ');
    x:= q[i, j];
```

```
    FOR k:= 1 TO n-1 DO
      x:= x + eqns[k, n+1]*p[i, j, k];
    writeln(x:10);
    IF x > max THEN BEGIN max:= x;  npy[i]:= j END
  END
END;
writeln(output);   writeln('New policy');
FOR i:= 1 TO n DO write(npy[i]:4);
writeln(output);   writeln(output);
FOR i:= 1 TO n DO
IF npy[i] <> py[i] THEN
BEGIN  py:= npy;  GOTO 1  END;
writeln('New policy same as old, so optimal')
END.
```

A print-out from the program, running on Dreyfus' example, is as follows:

```
Value-determination opn. for policy:
   1   1   1

Coeffts. of eqns. to be solved:
 5.000E-01  -2.500E-01   1.000E+00     8.000E+00
-5.000E-01   1.000E+00   1.000E+00     1.600E+01
-2.500E-01  -2.500E-01   1.000E+00     7.000E+00

Values of v are as follows:
 1.333E+00
 7.467E+00
         0
Average profit/trip 9.200E+00


Policy-Improvement Routine

State  1 altrntv.  1    1.053E+01
State  1 altrntv.  2    8.433E+00
State  1 altrntv.  3    5.517E+00
State  2 altrntv.  1    1.667E+01
State  2 altrntv.  2    2.162E+01
State  2 altrntv.  3    1.333E+00
State  3 altrntv.  1    9.200E+00
State  3 altrntv.  2    9.767E+00
State  3 altrntv.  3    5.967E+00

New policy
   1   2   2
```

```
Value-determination opn. for policy:
   1   2   2

Coeffts. of eqns. to be solved:
 5.000E-01  -2.500E-01   1.000E+00     8.000E+00
-6.250E-02   1.250E-01   1.000E+00     1.500E+01
-1.250E-01  -7.500E-01   1.000E+00     4.000E+00

Values of v are as follows:
-3.879E+00
 1.285E+01
         0
Average profit/trip 1.315E+01


Policy-Improvement Routine

State  1 altrntv.  1   9.273E+00
State  1 altrntv.  2   1.214E+01
State  1 altrntv.  3   4.886E+00
State  2 altrntv.  1   1.406E+01
State  2 altrntv.  2   2.600E+01
State  2 altrntv.  3  -3.879E+00
State  3 altrntv.  1   9.242E+00
State  3 altrntv.  2   1.315E+01
State  3 altrntv.  3   2.394E+00

New policy
   2   2   2


Value-determination opn. for policy:
   2   2   2

Coeffts. of eqns. to be solved:
 9.375E-01  -7.500E-01   1.000E+00     2.750E+00
-6.250E-02   1.250E-01   1.000E+00     1.500E+01
-1.250E-01  -7.500E-01   1.000E+00     4.000E+00

Values of v are as follows:
-1.176E+00
 1.266E+01
         0
Average profit/trip 1.334E+01
```

```
Policy-Improvement Routine

State  1 altrntv.  1    1.058E+01
State  1 altrntv.  2    1.217E+01
State  1 altrntv.  3    5.538E+00
State  2 altrntv.  1    1.541E+01
State  2 altrntv.  2    2.600E+01
State  2 altrntv.  3   -1.176E+00
State  3 altrntv.  1    9.870E+00
State  3 altrntv.  2    1.334E+01
State  3 altrntv.  3    4.409E+00

New policy
   2   2   2

New policy same as old, so optimal
```

The method can of course be used where the number of states, and number of alternative actions, are much greater. As already mentioned, the valuable characteristic of Dynamic Programming methods is that the computation time does not increase steeply with increase in problem complexity.

SIMULTANEOUS EQUATIONS

In the program for the Taxicab Problem, PROCEDURE solve was used without explanation. It in fact uses the method of elimination, or pivotal condensation, as explained in various books on Numerical Methods such as the well-known one from the National Physical Laboratory [7]. Methods for inversion of matrices are easily adapted to solving simultaneous equations, and some implementations of the BASIC language have built-in facilities for matrix manipulation.

REFERENCES

1. R.E. Bellman, **Dynamic Programming** (Princeton University Press, 1957)

2. R.J. Thierauf, **An Introductory Approach to Operations Research** (Wiley, Santa Barbara 1978) pp. 260-263

3. D. Pedoe, **The Gentle Art of Mathematics** (Penguin, Harmondsworth, 1963) pp. 55-57

4. M.G. Simpson, "Dynamic programming", in R.T. Eddison, K. Pennycuick and B.H.P. Rivett (eds.) **Operational Research in Management** (English Universities Press, London, 1962) Chapter 12, pp. 239-257

5. P.G. Carlson, **Quantitative Methods for Managers** (Harper & Row, New York, 1967)

6. S. Dreyfus, "Dynamic Programming", in R.L. Ackoff (ed.) **Progress in Operations Research**, vol. 1 (Wiley, 1961) Chapter 5, pp. 211-242

7. National Physical Laboratory, **Modern Computing Methods** (Notes on Applied Science No. 16, H.M.S.O., London, 1957) pp. 4-6

Chapter 8

CONTINUOUS SIMULATION

Up to now the emphasis has been on techniques which can be set up to search automatically for an optimum. It is not always feasible to analyse the situation to allow automatic search. Sometimes optimal, or at least acceptable, solutions must be sought by experiment.

It is often impossible, or at least highly dangerous, expensive, and/or time-consuming, to carry out experiments directly on the system it is desired to control. The system may be the economy of a country, or the operation of traffic lights at a busy junction, or the control of a chemical plant not yet built. If a computer program can be set up to simulate the system to be studied, the objections to experimentation are overcome, though of course the value of the results depends on the validity of the simulation.

Techniques for computer simulation tend to be discussed under two distinct headings. They may be termed 'continuous' and 'discrete-event' simulation. Systems to be simulated do not always fit neatly under one of these; it is sometimes desirable to combine continuous and discrete techniques in the same model. Some workers, for example Callaway [1], have prepared facilities allowing such combination. However, for simplicity, the two classes of simulation will be treated separately here, the present chapter being concerned with the continuous kind.

Continuous systems are the main objects of study of Control Engineering, and the use of analog computers to simulate dynamic systems is well known. The systems to be simulated are normally represented by sets of simultaneous differential equations. Such representation is also useful in connection with economic and ecological systems; for a time 'World Models' were popular as a means of trying to predict the future of the earth in terms of population, resources, pollution, and so on. The construction of these mammoth simulations is less popular now, presumably because it has been realised that there are unpredictable discontinuous changes in the human situation. Nevertheless, continuous models of social and ecological systems, on a somewhat more modest scale, are found to be valuable - see for example, Hall and Day [2]. The World Model idea is largely due to the work of Forrester [3] and Meadows [4]. Aspects of it have been criticised by Julien and Freeman [5].

It is not difficult to see that social and ecological systems can be described by sets of differential equations, with time as the independent variable. Expressions can be written down for the rate of exhaustion of a resource, or of growth of a population, or of the accumulation or depletion of the stock of a commodity. Many of the papers in Hall and Day show complicated arrow diagrams in which arrows indicate the influence of one variable on the rate of change of another.

DYNAMIC MODELS

In a simulation, the current state of the system being modelled is represented by values of a set of state variables. For simulation, by either analog or digital computer, it is convenient to express the relationships as a set of first-order differential equations. These allow the derivatives of state variables to be computed as functions of the current values of all the variables including the independent variable time. Formally, the equations are of the form:

$$dy_i/dt = f_i(t, y_1, y_2, y_3 \ldots\ldots)$$

where the y_i are the state variables.

The equations can be restricted to first-order without loss of generality. Since new state variables may be introduced as required, higher-order effects can be represented by sets of first-order equations. A simple example of a dynamic system to be simulated is an object satisfying the conditions which result in simple harmonic motion. That is to say, the object moves in a straight line and has an acceleration towards a point in the line, the amount of the acceleration being proportional to the distance from the point. The motion is represented by the second-order differential equation:

$$d^2x/dt^2 = -mu.x$$

in which mu is a constant. The equation can be broken down into two first-order ones by introducing the variable v to represent velocity:

$$dx/dt = v$$

$$dv/dt = -mu.x$$

or, using y_1 and y_2 to represent state variables, as:

$$dy_1/dt = f_1(t, y_1, y_2) = y_2$$

$$dy_2/dt = f_2(t, y_1, y_2) = -mu.y_1$$

INTEGRATION METHODS

Running a simultion of a model formed as above is exactly equivalent to obtaining a numerical solution to the set of differential equations. It is necessary to specify starting conditions. For the body obeying equations as above, initial values of the displacement y_1, and the velocity y_2 have to be specified. An initial value of the independent variable t is also needed, but is often assumed to be zero at the start of the run.

The simulation has to proceed in small increments of simulation time, say of magnitude h. At the starting time, say time 0, it is possible to compute the derivatives by evaluating the above functions $f_1(t, y_1, y_2)$ and $f_2(t, y_1, y_2)$. Strictly speaking, these derivatives are only correct for the time t = 0 for which they were evaluated, but if h is small it is reasonable to assume the derivatives are not greatly different over the interval from time 0 to time h. The new values of the state variables at the end of this interval can therefore be calculated approximately as:

$$y_1^1 = y_1 + h.f_1$$

$$y_2^1 = y_2 + h.f_2$$

These new values can again be used to evaluate the derivatives, and the new derivative values can be assumed to hold from t = h to t = 2h, and so on. In effect, the trajectories of the state variables are approximated by a series of straight-line segments, a segment going from time 0 to h, another from h to 2h, and so on. The accuracy of the simulation increases as h is reduced in magnitude.

The correct value of a state variable y_i at the end of one of the time intervals can be expressed as follows using Taylor's Theorem:

$$y_i(t + h) = y_i + h.y_i' + (h^2/2!)y_i'' + \ldots$$

in which y_i' is the first derivative, y_i'' the second, and so on. The method which was outlined, whereby the trajectories are approximated by straight-line segments, introduces an error due to the fact that the term in h^2, and all terms of higher degree in h, are omitted. When h is reduced to improve the accuracy, the neglected term which decreases most slowly is that in h^2. For small h, therefore, the error per integration step is proportional to h^2.

IMPROVING THE ACCURACY

If it can be arranged that the error per step is proportional to a higher power of h, the improvement in accuracy as h is reduced is much greater. A number of techniques have been devised in order to achieve this result. The most convenient ones to use belong to the class of methods associated with the names of Runge and Kutta. They are convenient because, as will be seen, there is no need to evaluate derivatives outside the time-interval under consideration. Some other methods can be shown to give a slightly better trade-off of accuracy against computational complexity, but as they require earlier values of the state variables they cannot be used at the start of the interval of time over which the solution is required.

The method described above could be termed a Runge-Kutta first-order process. It matches the terms in the Taylor expansion up to that of first order in h. A Runge-Kutta process of higher order requires a greater number of evaluations of the derivatives in each step, but it matches more terms of the Taylor expansion. A fourth-order process is often employed, requiring four evaluations per step and matching the terms of the expansion up to that of degree four, so that the error per step is proportional to h^5. Unless all that is wanted is a very crude simulation, it is well worthwhile to use the fourth-order process rather than the simple first-order one.

RUNGE-KUTTA METHOD

In the fourth-order method the derivatives are first evaluated, as before, at the start of the time interval. If t_0 is the time at the start of the interval, and y_1, y_2 ... the values of the state variables, the derivative of the i-th variable is $f_i(t_0, y_1, y_2 \ldots)$, and the increment which would be made in the first-order process is h times this amount. Let k_{i0} be the increment, which is formally evaluated as:

$$k_{i0} = h.f_i(t_0, y_1, y_2, \ldots)$$

In the fourth-order method a fresh evaluation of the derivatives is made for values of the time and other variables arrived at by allowing a fraction m of the above increment. Then a fresh set of increments, termed k_{i1}, are formed on the basis of these evaluations, and so on for two further stages.

$$k_{i1} = h.f_i(t_0+m.h, y_1+m.k_{10}, y_2+m.k_{20}, \ldots)$$

Then another set of values k_{i2} is obtained by combining a fraction (n-r) of the k_{i0} with a fraction r of the k_{i1}:

$$k_{i2} = h.f_i(t_0+n.h,\ y_1+[n-r].k_{10}+r.k_{11},\ y_2+[n-r].k_{20}+r.k_{21}, \ldots)$$

and then another set still by combining fractions (p-s-t), s and t of the earlier values:

$$k_{i3} = h.f_i(t_0+p.h,\ y_1+[p-s-t].k_{10}+s.k_{11}+t.k_{12},\ y_2+[p-s-t].k_{20}+s.k_{21}+t.k_{22}, \ldots)$$

and finally the increment in y_i is made by combining the k-values, as:

$$y_i^1 = y_i + a.k_{i0} + b.k_{i1} + c.k_{i2} + d.k_{i3}$$

where

$$a + b + c + d = 1$$

So far, the process is rather general, and a great many variables have been introduced without assigning values. By some rather tedious algebraic manipulation the increment in y_i can be expanded as a power series in h, and then the coefficients of the terms, up to that of fourth order, can be equated to those of the corresponding terms of the Taylor expansion. This produces a set of simultaneous equations in the variables m, n, r, p, s, t, a, b, c and d. The equations do not determine the variables uniquely. Any set of values satisfying the equations makes a version of the Runge-Kutta fourth-order process possible.

Gill [6] derives a set of values satisfying the requirements, and also achieving economy of storage when the program is run. The economy results from the fact that some of the numerical values in the computation need not be stored as they are easily computed from others. However, if economy of information storage is not of great importance, a solution which is more convenient to use is the following (see [7]):

$$m = n = r = 1/2; \quad p = t = 1; \quad s = 0$$

$$a = d = 1/6; \quad b = c = 1/3$$

so that the computations can be represented as follows:

$$k_{i0} = h.f_i(t_0, y_1, y_2 \ldots)$$

$$k_{i1} = h.f_i(t_0+(1/2)h, y_1+(1/2)k_{10}, y_2+(1/2)k_{20} \ldots)$$

$$k_{i2} = h.f_i(t_0+(1/2)h, y_1+(1/2)k_{11}, y_2+(1/2)k_{21} \ldots)$$

$$k_{i3} = h.f_i(t_0+h, y_1+k_{12}, y_2+k_{22} \ldots)$$

$$y_i^1 = y_i + (1/6)(k_{i0} + 2.k_{i1} + 2.k_{i2} + k_{i3})$$

One of the difficulties of using the method is that of choosing a suitable value for the interval h. If the value is too large the result is inaccurate, but an unduly small value means that the computation takes longer than it need. It is difficult to be sure that a suitable value has been chosen. A variant of the method due to Merson (see Fox [8] and Lukehart [9]) allows the automatic adjustment of the interval h during a run, according to a prearranged error criterion. The method can be programmed so that the interval is increased automatically if the precision is higher than is needed, or reduced if it is insufficient.

COMPUTER PROGRAM

A computer program implementing the method is as follows. An array called y holds the current values of the state variables, and it is convenient to let the independent variable, time, be held as y[0]. The main part of the process is carried out by PROCEDURE runkut.

The first parameter of PROCEDURE runkut is an integer n, indicating the number of dependent variables, and then a real h indicating the size of the integration step. The next parameter is an array called y which is made to correspond, in the call of the procedure, to the previously-declared array of the same name. This parameter is declared as a VAR parameter, so that the y-values are available outside the procedure, in case this is required.

The next two parameters are 'prnt' and 'fini'. The first determines the interval between print-outs and the second the duration of the run. To avoid possible difficulty from rounding-off errors they are both integers, representing the time intervals in units of the integration interval h.

The final parameter of PROCEDURE runkut is something not seen before. The parameter is another procedure, referred to by the name 'diff' within the procedure body. In the call of PROCEDURE runkut, the actual PROCEDURE shm (for simple harmonic motion) is substituted for 'diff'. The arrangement of parameters has to agree between 'diff' and whatever replaces it. Because 'runkut' has a procedure as parameter, it can readily be applied to any other set of simultaneous differential equations of the same first-order type (up to ten in number, as programmed here). All that is needed is to write a procedure corresponding to 'shm' for the other set of equations.

The program is as follows:

```
PROGRAM runge(input, output);

TYPE  vararr= ARRAY[0..10] OF real;
VAR   finish, h, pntint, mu: real;
      y: vararr;

PROCEDURE runkut(n: integer;  h: real;
    VAR y: vararr;  prnt, fini: integer;
    PROCEDURE diff(y: vararr; VAR f: vararr));
VAR i, j, stage, nn: integer;
    f, yy: vararr;
    k: ARRAY[0..10,0..3] OF real;
BEGIN  nn:= n;  IF n > 5 THEN nn:= 5;
  FOR i:= 0 TO fini DO
```

```
  BEGIN  yy:= y;
    IF i=prnt*(i DIV prnt) THEN
    BEGIN writeln(output);
      FOR j:= 0 TO nn DO write(y[j]:10,'   ')
    END;
    FOR stage:= 0 TO 3 DO
    BEGIN  diff(y, f);
      k[0, stage]:= h;
      FOR j:= 1 TO n DO k[j, stage]:= h*f[j];
      FOR j:= 0 TO n DO
      CASE stage OF
      0, 1: y[j]:= yy[j] + k[j, stage]/2;
      2: y[j]:= yy[j] + k[j, 2];
      3: y[j]:= yy[j] +
           (k[j,0] + 2*(k[j,1] + k[j,2]) + k[j,3])/6;
      END {case}
    END {loop in stage}
  END {loop in i}
END {runkut};

PROCEDURE shm(y: vararr; VAR f: vararr);
BEGIN f[1]:= y[2];  f[2]:= -mu*y[1]  END {shm};

BEGIN  h:= 0.1; {integration interval}
  mu:= 0.394784; {acceln. constant}
  y[0]:= 0; {start time}
  y[1]:= 1;  y[2]:= 0; {initial conditions:
                  displacement = 1, velocity 0}
  finish:= 10; {finish time}
  pntint:= 1; {print interval}
  runkut(2, h, y, {2 is no. of eqns.}
    round(pntint/h),  round((finish-y[0])/h),  shm)
END.
```

and the output from a run as follows:

0.000E+00	1.000E+00	0.000E+00
1.000E+00	8.090E-01	-3.693E-01
2.000E+00	3.090E-01	-5.976E-01
3.000E+00	-3.090E-01	-5.976E-01
4.000E+00	-8.090E-01	-3.693E-01
5.000E+00	-1.000E+00	-7.562E-07
6.000E+00	-8.090E-01	3.693E-01
7.000E+00	-3.090E-01	5.976E-01
8.000E+00	3.090E-01	5.976E-01
9.000E+00	8.090E-01	3.693E-01
1.000E+01	1.000E+00	1.319E-06

The values have been plotted in Figure 8.1, and

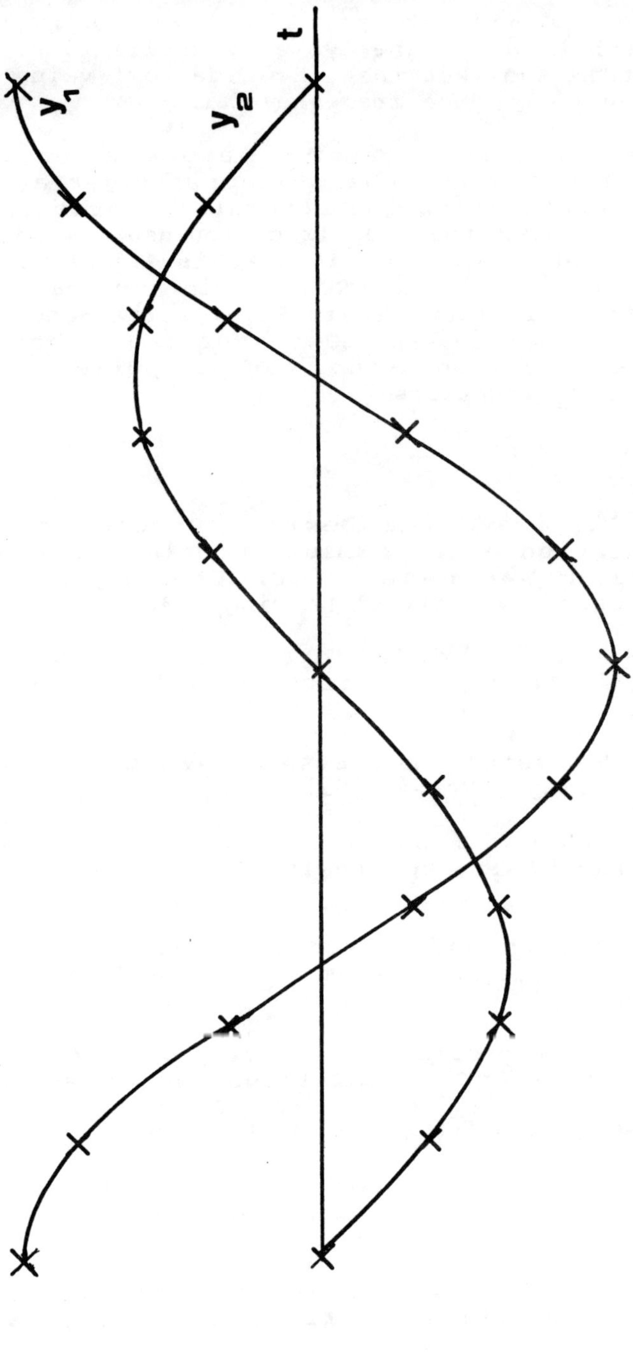

Figure 8.1. Plot (as crosses) of the output of the Runge-Kutta program.

it can be seen that the period of the simulation is just enough to allow one cycle of oscillation. This results from the fact that the numerical value of mu was chosen to produce frequency 0.1.

The simple-harmonic motion example is, of course, a particularly simple one. An amusing variation is made by substituting the accurate equations for a pendulum, without the simplification usually made for low-amplitude oscillations. It is found that the period of the oscillation differs in accordance with the theorerical treatment in Physics text-books (for example Watson [10]). Once again it should be emphasised that the method can be applied to much more complex situations.

REFERENCES

1. A.A. Callaway, **The Design, Development and Implementation of a Combined Continuous-Discrete Simulation Language** Ph.D. Thesis, Dept. of Cybernetics, University of Reading, 1975

2. C.A.S. Hall and J.W. Day (eds.) **Ecosystem Modeling in Theory and Practice** (Wiley, New York, 1977)

3. J.W. Forrester, **Industrial Dynamics** (M.I.T. Press, Cambridge, Mass. 1961)

4. D.H. Meadows, D.L. Meadows, J. Randers and W.W. Behrens, **The Limits to Growth** (Earth Island, London, 1972)

5. P.-A. Julien and C. Freeman, "The Capital and Industrial Output Sub-System." **Futures 5**, no. 1, pp. 66-79, Feb. 1973

6. S. Gill, "A process for the step-by-step integration of differential equations in an automatic digital computing machine" **Proc. Cambridge Philosophical Society 47** pp. 96-108, 1951

7. National Physical Laboratory, **Modern Computing Methods** (Notes on Applied Science No. 16, H.M.S.O. London 1957), p. 59; or 2nd Edition, 1961, pp. 87-88

8. L. Fox (Ed.) **Numerical Solution of Ordinary and Partial Differential Equations** (Pergamon, Oxford, 1962) p. 24

9. P.M. Lukehart, Algorithm 218 Kutta Merson, **Comm. of the A.C.M. 6**, no. 12, pp. 737-738, Dec. 1963

10. W. Watson, **A Text-Book of Physics** (Longmans, Green, London, eighth edition 1923 and 1941) p. 126

Topics such as the Simple Pendulum tend to be treated in more detail in older books, such as Watson's. A recent publication which mentions the effect of amplitude on period is:

D. Humphrey and J. Topping, **A Shorter Intermediate Mechanics** SI Edition (Longmans, London, 1971) pp. 300-301

Chapter 9

MONTE-CARLO METHOD

Simulations of continuous systems, as discussed in the last Chapter, are normally deterministic. That is to say, once the simulation begins from a particular set of starting conditions, what happens to it thereafter is effectively pre-determined.

It is also useful to simulate systems having probabilistic, or stochastic, properties. The 'taxi-cab' situation described in connection with Dynamic Programming was stochastic, since it was impossible to say for certain what state transition would follow an action chosen by the driver.

Sometimes stochastic situations can be analysed without recourse to simulation. The taxicab problem is an example. A very large body of mathematical theory is available to aid such analysis. If simulation is needed, the computer must be persuaded to behave unpredictably, or at least in a way which makes it difficult to predict.

The nature of a digital computer is such that (if it is in good working order) its operations are totally deterministic. It could, of course, be connected to a special device to generate random inputs from, for example, resistor noise or radio-active decay. Some computers have been built incorporating such a device. Their instruction sets included one which introduced random numbers from the truly-random source.

Modern computers do not usually include a means of generating truly-random numbers. It is found that the same purpose is served by special program routines called 'pseudo-random number generators'. This is often abbreviated to p.r.n.g.

The term 'Monte Carlo Method' is used to refer generally to methods for solving problems, depending on random (or pseudo-random) sampling. The introduction here has been in connection with simulation, but sometimes the method is presented as a means of solving purely mathematical problems. If a mathematical problem can be linked to a physical model, the result of running a simulation of the physical system can be presented either as data on it or as a solution of the mathematical problem.

Page [1] has defined the Monte-Carlo Method as a means of solving mathematical problems, as follows:

> "Briefly the method consists of finding a stochastic process whose analytical solution can be made to depend on the mathematical problem it is desired to solve, performing artificial sampling experiments to obtain realisations of the stochastic process, and then interpreting the results of the experiments to provide a numerical solution of the mathematical problem."

Although Page's primary concern is with the mathematics, it is clear that the method is also applicable where the main interest is in the physical stochastic process itself. He goes on to say that the method can be used to evaluate multi-dimensional integrals, to invert matrices and to solve some types of equation. These include linear, partial differential and integral equations.

The early use of the Monte-Carlo Method was in connection with Atomic Physics. It is frequently necessary to know the distribution of atomic particles emerging from some medium in which they make repeated collisions. It may be impossible to solve the problem analytically, especially if the boundaries of the medium do not conform to some simple shape such as a sphere or cylinder or infinite slab. Using the Monte-Carlo Method it is possible to trace the path of a simulated particle through the medium. The pseudo-random (or random) number generator is invoked each time the particle makes a collision and determines the new direction of travel, and perhaps the free path till the next collision.

By simulating the paths of a great many particles through the medium, the spatial distribution of emerging particles can be approximately inferred. It is of course necessary to arrange that the pseudo-random numbers are samples from distributions suitable to the physical situation.

Monte-Carlo simulations were carried out by atomic physicists even before digital computers were readily available, using such well-known physical effects as the rolling of dice to generate the random numbers. The procedure must have been extremely tedious.

The first integral equation discussed by Page is one which arises in a problem of finding the distribution of particles making repeated collisions along a line (which could be through the thickness of a slab). This discussion illustrates very well the use of the Monte-Carlo Method in connection with Atomic Physics. It is also an impressive demonstration of how a formidable-looking mathematical problem can be solved quite simply by the use of the method. Only a numerical approximation, not an analytical solution is produced, but it is adequate for practical purposes.

THE BUFFON NEEDLE PROBLEM

An intriguing use of the Monte-Carlo Method is for the evaluation of 'pi'. It does not lend itself to precise evaluation because of physical limitations which will become apparent, but it is interesting that the idea dates from 1760 (see Pedoe [2], Novikoff [3] and Kemeny and Kurtz [4]).

For the Needle Problem it is necessary to imagine a flat surface ruled with a regular grid of lines with a spacing of d. Suppose a needle of length l, where $l \leq d$, is tossed randomly onto the grid. It can be shown that the probability that the needle intersects a line of the grid is (2/pi)(l/d).

If it is assumed for simplicity that l = d, the probability of an intersection is (2/pi). If the needle is tossed on to the grid a great many times, this should be approximately the proportion of throws which result in the needle intersecting a line. It is therefore possible to estimate pi, the ratio of the circumference of a circle to its diameter, as:

2(Total no. of throws)/(No. of intersections)

Needless to say, this is not a very good way to estimate pi, though Pedoe refers to an experiment in which the value was obtained to six decimal places. The procedure is interesting as an early example of the method.

PSEUDO-RANDOM NUMBER GENERATORS (p.r.n.g.)

A short survey of the main kinds of p.r.n.g. is given by Andrew [5]. Since they are implemented on deterministic machines they are themselves deterministic, and any sequence of numbers produced is entirely determined by the initial state of the generator. An integer, sometimes referred to as a 'seed', sets the initial state. Sometimes it is arranged that the seed is set automatically from the time of day, and perhaps the date, at which the program is run. When this trick is used it may appear to a user that the generator is truly random since the sequence is different in successive runs. All of the methods generate numbers cyclically, but for the well-known methods the length of the cycle is sufficient to ensure that the numbers will almost certainly not begin to repeat in any one simulation run.

Statisticians often use tables of random numbers, and instead of programming a p.r.n.g. it would be possible to store such a table in one of the forms of secondary storage (magnetic disk or tape, say) of the computer. Instead of calling the p.r.n.g., numbers would be taken in sequence from the table. The two methods are actually very similar and a p.r.n.g. is really a table held in an extremely compact form. It is, of course, necessary that the sequence of numbers, whether in a table as such or generated by a p.r.n.g., be subjected to statistical tests of randomness. There is much debate as to just what these should be. It can probably be assumed that any widely-used p.r.n.g. has not given a disastrous result in any of the tests.

Mainly because they use modular arithmetic, p.r.n.g.'s are usually programmed in machine language. The most widely-used type of generator is the multiplicative one, in which each number in the sequence is obtained by multiplying the one before it by a constant. The multiplication is carried out in modular arithmetic; that is to say, the effective result is not the full result of the multiplication but the remainder on dividing it by a number M called the modulus. If the successive numbers are u_1, u_2, u_3 ... and the multiplier is k, this method may be represented by:

$$u_{n+1} = k.u_n \pmod{M}$$

Probably the most widely-used p.r.n.g. is the multiplicative one in which:

$$k = 455,470,314 = 13^{13} \pmod{M}$$

$$M = 2^{31} - 1$$

$$0 < u_n < M$$

For this generator, every number satisfying the constraint on u_n appears in the cycle, so any number in this range can form the starting seed and the generator will enter the long cycle. The cycle length is therefore (M-2) or just over two thousand million numbers.

Normally the generator is programmed in machine language. One reason is that it is necessary to allow for very large integers appearing after the multiplication is done, but before the result is reduced to be modulus M. Even this result may be inconvenient to store since 31 binary digits are needed. A multiplicative generator which is sometimes used is one in which:

M = any power of 2, say 2^{20} or 1,048,576

k = any odd power of 5 such that k < M, say 5^5 or 3125

u_0 = any odd positive integer < M

Cycle length = M/4

(M/4, since half of the odd numbers are in each of the two possible cycles)

The SPECTRUM can store integers up to $(2^{32}-1)$ with precision, so this generator can be programmed as follows as a subroutine in line 1000:

```
1000   LET  seed = seed*3125:
       LET  seed = seed - 1048576*INT(seed/1048576):
       LET  ran  = seed/1048576: RETURN
```

in which 'ran' is the output of the p.r.n.g., adjusted to be in the range 0 to 1. It is usual to adjust the output to be in this range. The value of 'seed' must be set to an odd positive integer value before the subroutine is first called.

A program embodying this required 8608 seconds to generate all the 262,144 numbers in the cycle, so the rate of generation is about 30 per second.

In practice it is hardly ever necessary to program a p.r.n.g. unless to meet some special requirement. Most implementations of high-level languages have a built-in p.r.n.g. In SPECTRUM Basic a pseudo-random value is obtained from the standard function RND. A RANDOMIZE statement can be used if desired to set the initial state of the p.r.n.g., and if it is of the form RANDOMIZE 0 (or RANDOMIZE on its own without any number following) the effect is to initialise the p.r.n.g. from the real-time clock.

The RND function generates numbers a little faster than does the home-made p.r.n.g. listed above. RND can generate about 47 numbers per second. Runs of Monte-Carlo programs often require the generation of a great many pseudo-random values, so the overall running time depends very much on the time needed by the p.r.n.g.

In the version of PASCAL used here, a pseudo-random output is given by the FUNCTION rand, which must first be declared as EXTERNAL (see the next sample program).

SAMPLES FROM OTHER DISTRIBUTIONS

The output from a p.r.n.g. normally takes all possible values over a particular range of integers. Of the two types of multiplicative generator which have been seen, the first produces all possible values over a particular range. The other is rather unusual in having two alternative cycles in which it may operate, but in either of them the values produced are regularly spread over a particular interval.

One of the statistical requirements for a p.r.n.g. is that there is much the same distribution of numbers in any reasonably-large sequence of generated values, irrespective of where in the cycle the seed value initiates the sequence. The distribution of numbers must therefore be similar to the distribution over the whole cycle. In other words, it has to be a rectangular distribution in which the probability of a value in the interval x to x+dx is proportional to dx, provided this interval falls within the permitted range. As already seen, it is usual to scale the output of a p.r.n.g. so that the permitted range is from 0 to 1.

What is provided by a basic p.r.n.g., then, is samples from a rectangular distribution over the range 0 to 1. For some purposes it is necessary to produce numbers which are samples from some other known distribution. These can be produced by operating suitably on the output values from the p.r.n.g.; techniques for doing it are discussed fully by Tocher [6].

Samples from the Gaussian or normal distribution, as in Figure 9.1(a), are often wanted. A simple way of generating samples from a distribution of approximately this form is to add together a large number of samples drawn successively from the rectangular distribution. The Central Limit Theorem assures us that the distribution must approach the Gaussian form as the number of independent samples is increased. In practice, a pretty good approximation is formed by adding batches of twelve numbers from the basic p.r.n.g. to form each number. The mean value of 6 should be subtracted, in order to let the new distribution be centred on zero (or alternatively, 0.5 could be deducted from each of the twelve numbers before the addition).

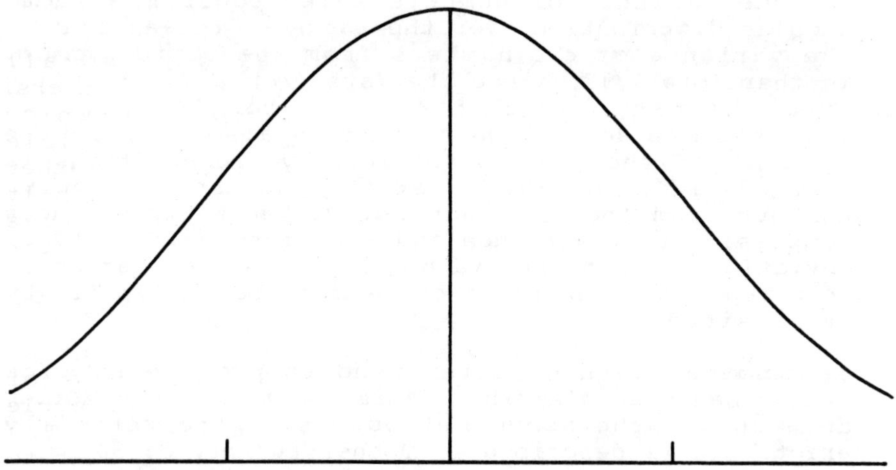

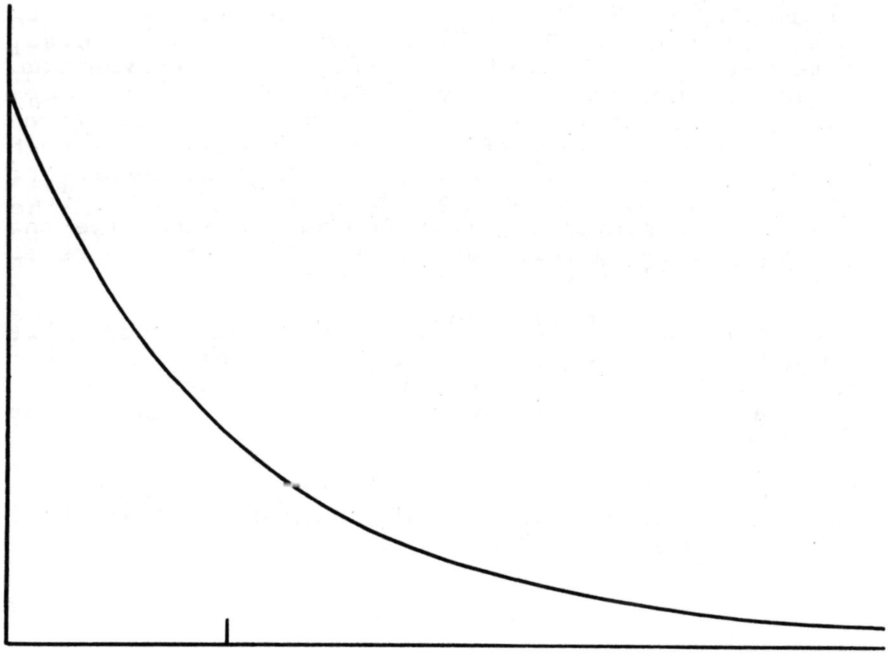

(b)

Figure 9.1. (a) Gaussian or normal distribution. (b) Negative-exponential distribution.

The variance of numbers selected from a rectangular distribution over the range -a to +a is $a^2/3$. The variance of the numbers from the basic p.r.n.g. is therefore 1/12, since they are from a distribution over the range -1/2 to +1/2. When independent variates are added, the variances also add, so the variance of the numbers obtained by the additions is 12 times 1/12, or unity. So the addition of twelve numbers from the basic p.r.n.g. conveniently produces samples of unit variance and therefore unit standard deviation. If another value of standard deviation is required, each generated value must be multiplied by the desired s.d.

Another method often used to produce samples from a Gaussian distribution is due to Box and Müller and, unlike the above method, it is theoretically exact. It is described in Tocher [6] pp. 33-34.

Samples from a negative-exponential distribution, as in Figure 9.1(b), are also often wanted. When events happen in Poisson fashion, the inter-event intervals have a negative-exponential distribution. Events happen in this way if each one occurs independently of the others; the atomic disintegrations in radioactive decay are an ideal example. Although it is easy to think of reasons why the behaviour should not be precisely of this nature, it is often assumed for simplicity that sporadic events like the arrival of emergencies at a hospital or customers at a bank, have a Poisson distribution.

Where a discrete-event simulation (see next Chapter) involves one or more sequences of events happening in Poisson fashion, it is necessary to generate pseudo-random values for the inter-event intervals.

If r is the rate, or mean number of events per unit time, the probability of an inter-event interval falling between t and t+dt is:

$$r.e^{-r.t}.dt$$

where e is the base of Napierian logarithms.

Numbers which are samples from this distribution are very easily generated. If x is a sample from a rectangular distribution over the range 0 to 1, the value computed as:

$$-(1/r)\ln(x)$$

is a sample from a negative-exponential distribution having an average rate of r events per unit time, or (1/r) as the mean inter-event interval.

A small practical point is that, if x can take the value zero, -ln(x) is then infinite and will produce an overflow condition in the computer. If, as is fairly usual, the smallest value which can be produced for x is not identically zero, no problem arises. If the value zero can arise, the simplest solution is to arrange to test for it and to replace it with a further sample from the p.r.n.g. when it appears.

The use of these methods will be illustrated in a program for discrete-event simulation, in the next Chapter. They illustrate two of the main ways of operating on the output of a p.r.n.g. in order to obtain samples from special distributions. One is by combining a number of outputs of the p.r.n.g. (here, it was by adding them, but they could have been multiplied, or their maximum value taken), and the other is by a non-linear transformation (such as taking the logarithm).

A third method which can be used is the 'rejection method' and it allows the generation of samples from an arbitrary distribution, provided it has value zero outside a certain finite range. Suppose that, over the range from A to B, the distribution is represented by a function f(x). Let M be the maximum value of f(x).

To generate a number by the rejection method, two successive numbers are taken from the p.r.n.g., say x and y. The first is scaled and incremented so that it is a sample from a rectangular distribution over the range A to B. The second is scaled so that it is a sample from a rectangular distribution over a range from 0 to a value not less than M. Then f(x) is evaluated. If y is less than f(x), the value x becomes an output. Otherwise, these two values are both rejected and a fresh pair is taken from the p.r.n.g., and the process is repeated until the condition for an output is satisfied.

The rejection method cannot be used directly if the range of variation is infinite. However, the three methods can be used in combination. The output from an implementation of the rejection method can be subjected to a nonlinear transformation, such as taking the logarithm, to produce samples from a distribution over an infinite range.

THE TAXICAB AGAIN

A problem faced by a hypothetical taxi driver was solved by Dynamic Programming in Chapter 7. If the method applied there had not been devised, the matter could have been investigated by Monte-Carlo simulation. The various possible policies can be evaluated one at a time. In the given example there are only nine of them, so it is not unduly tedious to evaluate them all. The great advantage of the Dynamic Programming method is that it remains manageable when the situation is much more complex.

Suppose it is required to evaluate the policy which was earlier found to be optimal, namely that where action no. 2 (go to a rank) is always chosen. The transition probabilities between states are then represented by the matrix:

1/16	3/4	3/16
1/16	7/8	1/16
1/8	3/4	1/8

which has to be inserted in decimal numbers as:

0.0625	0.75	0.1875
0.0625	0.875	0.0625
0.125	0.75	0.125

and the matrix of returns is:

8	2	4
8	16	8
6	4	2

Suppose the simulated system is in State 1. It is necessary to build into the program an arrangement which lets its next state be chosen from the three possibilities with probabilities as follows:

0.0625 0.75 0.1875

The simplest way to arrange this is to convert these to cumulative probabilities, as follows:

0.0625 0.8125 1.0

and then, when it is required to make a choice, a pseudo-random number is generated as a sample from the rectangular distribution over the range 0 to 1. If this sample, say x, is less than or equal to the first number, namely 0.0625, the next state of the system is chosen to be state 1. Otherwise, x is compared with the second number, and if it is less than or equal to it, the next state is state 2. If the last number is reached it must be chosen, and in the program a special provision is made for this. This should be unnecessary if the last number is indeed unity, but is probably advisable in case there are rounding-off errors.

A program to evaluate a fixed policy, written to allow for only three states of the system, is as follows. The simulation is made to start, arbitrarily, from state 1, and it runs for 1000 journeys of the taxicab.

```
PROGRAM taxisim(input, output);

LABEL 1;

VAR  p, r: ARRAY[1..3,1..3] OF real;
     i, j, state, count: integer;
     ret, x: real;

FUNCTION rand: real;  EXTERNAL;

BEGIN  writeln('Insert probty. matrix');
  FOR i:= 1 TO 3 DO FOR j:= 1 TO 3 DO
                   read(p[i, j]);
  writeln('Insert returns matrix');
  FOR i:= 1 TO 3 DO FOR j:= 1 TO 3 DO
                   read(r[i, j]);

  {Convert probabilities into cum. probts.}
```

```
  FOR i:= 1 TO 3 DO
  BEGIN  FOR j:= 2 TO 3 DO
    p[i, j]:= p[i, j] + p[i, j-1]
  END;

  ret:= 0;  state:= 1;
  FOR count:= 1 TO 1000 DO
  BEGIN  x:= rand;
    FOR j:= 1 TO 3 DO
    IF (j=3) OR (x <= p[state, j]) THEN
    BEGIN ret:=ret + r[state, j];
      state:= j;  GOTO 1
    END;
  1: END;
writeln('Average return per trip', ret/1000)
END.
```

When run with the transition and return matrices for the policy [2, 2, 2], as above, this program produced the output:

Average return per trip 1.3394001E+01

That is to say, the average return per trip estimated by simulation was 13.394, compared to 13.34 from the Dynamic Programming method.

For the policy [1, 1, 1] the average return was estimated by simulation as 9.136, whereas the Dynamic Programming method gave 9.200. It can be seen that the two methods are in fairly good agreeement.

REFERENCES

1. E.S. Page, "The Monte Carlo Solution of some Integral Equations." **Proc. Cambridge Philosophical Society 50**, pp. 414-425, 1954

2. D. Pedoe, **The Gentle Art of Mathematics** (Penguin, Harmondsworth, 1963), pp. 59-60

3. A.B.J. Novikoff, "Integral Geometry in Pattern Perception", in **Principles of Self-Organization** ed. H. von Foerster and G.W. Zopf (Pergamon, Oxford, 1962) pp. 347-368

4. J.G. Kemeny and T.E. Kurtz, **Basic Programming** (Wiley, New York, 2nd ed. 1971) pp. 68-69

5. A.M. Andrew, **Artificial Intelligence** (Abacus, Tunbridge Wells, 1983) pp. 16-20

6. K. Tocher, **The Art of Simulation** (English Universities Press, London 1963 and 1967)
{It should be noted there is a small error in the Box-Müller formula in the earlier edition.}

Chapter 10

DISCRETE-EVENT SIMULATION

In the continuous-variable type of simulation it is necessary to let simulation time proceed in very small steps referred to as 'integration intervals'. This is because the state variables are changing continuously. Sometimes, however, the system to be simulated is such that there is no change in any of the state variables (except time, itself) over substantial periods of time. If, at some point in the simulation, it is possible to predict the next event which will change the state variables, then the most efficient thing to do is to let the simulation time advance in one jump to the time of that event.

A simple discussion of the requirement is given by Williams [1]. The system to be analysed has to be represented in terms of two types of event, which he terms 'delayed events' and 'conditional events'. A delayed event is one whose occurrence can be predicted at some earlier point in simulation time. For example, if a doctor begins his period of consultation with a patient, it is possible to predict the finish of the consultation, and to assign a time to it. It is not necessary to assume that consultations are of uniform duration; the duration may be dependent on the output of a pseudo-random number generator as discussed in the last Chapter. From the point of view of the organisation of the simulation, however, the use of the p.r.n.g. to assign a duration is best done at the start of the consultation period. Once the duration has been assigned the finish-event is a delayed event.

If the simulation requires a sequence of events to happen in Poisson fashion, the inter-event intervals have a negative-expontial distribution. Whenever one of these events occurs, the p.r.n.g. can be used to decide the next inter-event interval and thus to predict the next event in the sequence. The arrival of patients at a surgery is likely to be assumed to be Poisson. There may be several such sequences of events running in parallel in a simulation.

To organise the simulation it is necessary, in effect, to set an alarm clock to ring when a delayed event must occur. When a number of alarm clocks are running at the same time, they must be scanned to find which is the next to ring.

The simulation must also include events which cannot be predicted, but which happen whenever some set of conditions is met. For example, the start of a consultation with a doctor (in a multi-doctor clinic) happens as soon as there is at least one patient waiting and at least one doctor free.

As soon as a delayed event has been simulated, a check must be made as to whether the conditions are now right for any of the possible types of conditional event. If they are, the event should be simulated. When the possibilities for such events have been exhausted, the alarm clocks should be scanned to find when the next delayed event is due, and simulation time advanced accordingly.

SIMULATION PACKAGES

The sequencing of events as described in the foregoing (involving scanning of alarm clocks) can be arranged, from scratch, in a program written in a standard language. However, the task of preparing the simulation program is much easier if certain facilities are available in a 'simulation package' or even a special simulation language. Williams describes a package with the name E.S.P. (for Elliott Simulation Package, not Extra-Sensory Perception) embedded in ALGOL 60. A PASCAL package giving the same facilities will be shown here.

These packages are very primitive as simulation packages go, but they illustrate quite adequately the use of delayed and conditional actions to determine the sequencing of events. Other packages embody the means of treating sets of objects (for example patients) passing through the simulation as distinguishable, rather than just as the count of total number. The E.S.P. package, and the PASCAL version here, readily allow the collection of certain statistics such as the totals of patients seen by each doctor. They do not lend themselves so conveniently to some other tasks, such as the collection of statistics about the times patients are kept waiting.

It is possible with some ingenuity to extend the simple packages to facilitate the collection of other statistical data. It can be arranged, for example, that each patient is assigned a serial number and that a record is kept of his time of arrival, so that statistics on waiting times can be obtained.

(Alternatively the result can be achieved rather more simply, but perhaps with less generality, by arranging the effect of a 'buffer', or 'first-in-first-out' store for the arrival times. If it is assumed that patients are seen in order of arrival, a patient's arrival time can be picked from the buffer when his turn comes, so his waiting time is simply the difference between this and the current simulation time.)

Simulation studies are useful in examining the likely effects of proposed changes in the simulated system, such as the installation of an extra facility (an extra X-ray machine, or the employment of an extra doctor, say), or the adoption of new administrative methods for steering patients through the system.

SAMPLE SIMULATION: MULTI-DOCTOR CLINIC

Williams uses the example of a clinic with four doctors which remains open for three hours. He treats the case where patients arrive regularly, and consultations require a standard amount of time. As presented here, the simulation has been modified to have stochastic characteristics, in that patient arrivals are Poisson, with an average inter-event time of 7 minutes. The time taken for a consultation is a sample from a Gaussian distribution with mean value 9 minutes and standard deviation 1 minute.

Initially, all four doctors are free. The doctors are numbered from 1 to 4, and where more than one doctor is free a patient goes to the lowest-numbered of those available. When the clinic door is closed after 180 minutes, no further patients are admitted but any waiting are seen.

COMPUTER PROGRAM

The program uses some features of PASCAL not previously introduced. Before discussing these, however, the facilities offered by the simulation package have to be introduced.

Most of the program is part of the package and can be transferred as it stands to other discrete-event simulations. The following variables have been introduced for this specific example - (open, patients, i, seen, docfree). Apart from the declarations of these variables, the only part of the program specific to this simulation is at the end, starting with the line:

```
open:= true;  patients:= 0;
```

A procedure and two functions are declared as part of the package. PROCEDURE call provides the means of setting an alarm clock to trigger the simulation of a delayed action. Types of delayed action are indicated by numbers, and of the two parameters of this procedure the first indicates the type of action to be triggered and the second is the amount of simulation time to elapse before the triggering happens.

FUNCTION next is the means by which the package controls the sequence of events. It delivers an integer value which indicates the type of the next delayed event, or else the value zero which initiates the check to see whether conditions are right for any of the conditional events. FUNCTION notdone is also provided to deliver a Boolean value which becomes 'false' when no alarm clocks are set and a final check of the conditions has been made.

To program the simulation the user must initialise some variables (all the entries in the array 'docfree' are set to 'true', and the entries in 'seen' are set to zero). Two calls of PROCEDURE call are made, to set the simulation in motion. The main part of the simulation is then embedded in the loop beginning:

```
WHILE notdone DO
CASE next OF
```

and finally, outside the loop, there is a statement to print out the results.

The situation to be simulated is represented by three kinds of delayed action, and one kind of conditional action. The three delayed actions are:

(1) Arrival of a patient.

(2) Finish of a consultation.

(3) Shutting the clinic door.

Conditional action: start of a consultation.

In the CASE statement within the loop written by the user, there are sections of program introduced by the numbers 1, 2, 3 and 0. The order of these in the program is unimportant and they indicate what must done in order to simulate each type of action. It can be seen that they invoke PROCEDURE call to set alarm clocks for further delayed actions.

A further point which has not been mentioned is that the alarm clocks have a facility for storing information and making it available when they ring. Two variables of type 'datarr' (data arrays) are declared, called 'get' and 'send'. Any numerical data held in 'send' when an alarm clock is set (by a call of PROCEDURE call) is stored until the respective delayed event is triggered, whereupon the data appears in 'get'.

In the section of program (introduced by 0:) dealing with the conditional event, the call of PROCEDURE call is preceded by an assignment to send[1] of the serial number of the doctor conducting the consultation. This makes it possible, at the end of the consultation (delayed action 2) to arrange that the appropriate doctor again becomes free. The statement arranging this is:

```
docfree[round(get[1])]:= true;
```

FUNCTION round is needed to convert the real value to integer form. For maximum flexiblity, it has been arranged that 'get' and 'send' are arrays of reals.

FUNCTION gaussian and FUNCTION negexp produce samples from, respectively, Gaussian and negative-exponential distributions. The Gaussian sample is produced with zero mean and unit standard deviation; if it is desired to have a sample with mean m and standard deviation sd, it has to be computed as:

```
m+sd*gaussian
```

Similarly, if the negative-exponential sample is to be from a distribution with mean m it has to be computed as:

```
m*negexp
```

In order to write the necessary additions for a particular simulation task it is not strictly necessary to understand the working of the package, but the following notes should make it clear.

Among the type declarations at the start of the program there is one of TYPE event. This is a variable in the form of a record to hold all the information which has to be stored in order to trigger a delayed action at some future time. The record has four fields. One is called 'tim' and holds the value of simulation time at which the action is to happen. The next is called 'no' and holds the type-number for the action (delayed action type 1, 2 or 3 in the present example). The next is called 'data' and holds the information which is salted away by the use of 'send' and becomes available through 'get'. The fourth field is called 'link' and is of TYPE pntevent, or pointer to an event.

In the explanation of the simulation principle, several references were made to scanning the alarm clocks to find which is the next to ring. In fact, the program does not operate quite as this would suggest. The information on the alarm clocks is stored in order of time of occurrence, so the next one to ring is simply that at the head of the list.

LIST PROCESSING

Variables of TYPE event contain pointers to other similar variables. The possibility of linking variables (always record types) in this way allows the construction of Data Structures and of what is referred to by the prosaic term List Processing. Where each record contains pointers to two or more others, the data structure can take the form of a tree. The variables of TYPE event contain only one pointer in each, so the only type of structure possible is a simple ordered list, as in Figure 10.1(a). The variable 'stack' of TYPE pntevent, points to the first 'event' in the list, that is, the one whose time of occurrence has the lowest numerical value. The 'link' field in this points to the 'event' with next lowest time value, and so on until a NIL pointer is reached.

Initially (and finally) the stack of events is empty, so the variable 'stack' has value NIL.

The use of list-processing techniques to form data structures, especially tree-structures, allows the processing of symbolic, rather than numerical, information. Programs can be written to simplify algebraic expressions, or to perform symbolic differentiation. A short introduction to the techniques and their use for symbolic differentiation is given by Andrew [2].

For the present application there are two advantages in using list-processing techniques rather than simply an array of variables of TYPE event:

(a) It is not necessary for the programmer to make a guess at the maximum size to which the list may grow during a run of the program. For an array, he has to make a guess in order to set the bounds. Of course, the list cannot ever grow without limit; the system as a whole is finite. However, if several lists were to be in use at the same time, the list-processing solution would have the advantage that they would all share the same area of storage, allowing some to become large provided the others remained small.

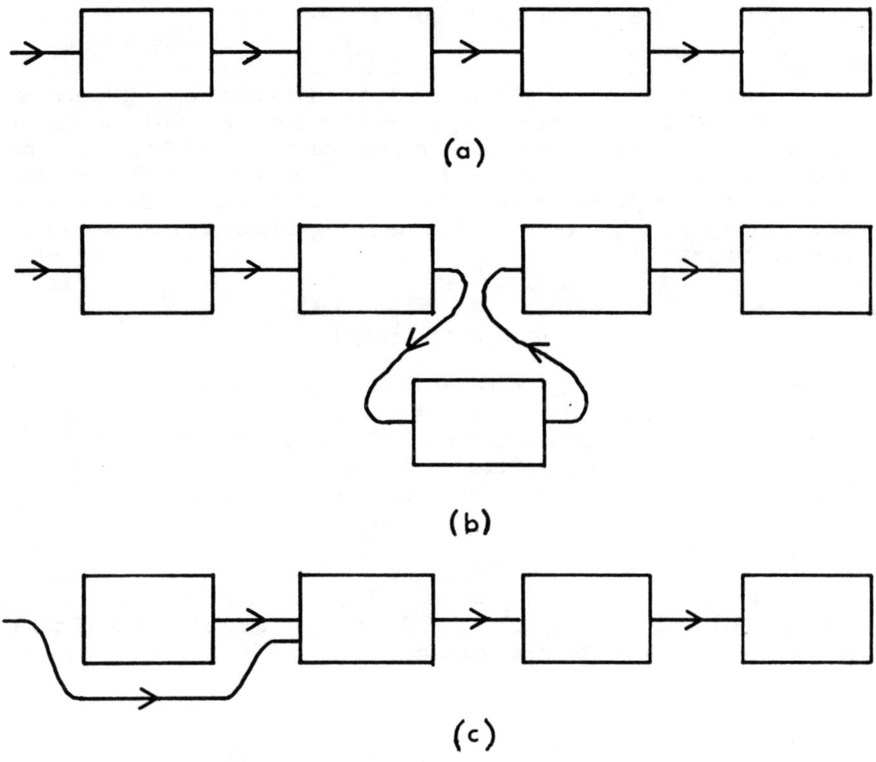

Figure 10.1. (a) List structure formed of records containing pointers to others. (b) Introduction of an element to the middle of a list. The 'link' from the record to the left of the insertion must be copied into the new one. (c) Removal of an element from the head of a list, by copying its 'link' into the pointer to the list.

(b) More important in the present application is the fact that items can be added to the list, at either end or in the middle, without the need to shift elements along to make room. Also, elements can be removed from the head end of the list without having to shift elements along to fill the space. Figures 10.1(b) and 10.1(c) illustrate these possibilities.

The hat sign (^) is used to indicate a pointer. (It should be an upward-pointing arrow, but because this is not included in most character sets, the arrow-head on its own has to suffice.) To designate a pointer-type, the hat is placed before the designation of the type to which it points, so that the declaration:

```
pntevent= ^event;
```

indicates that a 'pntevent' is a pointer to an 'event'. Then, where e is declared to be of TYPE pntevent, the combination e^ is used to indicate the event to which e points. The separate fields of the 'event' are accessed in the usual way, either by writing:

```
e^.tim
```

to designate the 'tim' field of the event to which e points, or else by writing:

```
WITH e^ DO
```

to introduce a section of program wihin which 'tim', is understood to mean e^.tim, and similarly for the other fields.

Within PROCEDURE call, the recursive PROCEDURE putin is made to run along the list until it finds the right place to insert the new 'event'. When it finds the place, its formal parameter e is a pointer to that part of the list which will be beyond the new element inserted. A copy of e is made in ee, so that it can later become the 'link' field of the new element. The new 'event' is generated by a call of PROCEDURE new, as:

```
new(e);
```

The fields of the new 'event' are then filled in appropriately - see Figure 10.1(b).

PROCEDURE next operates in two different ways according to whether it finds 'checked' to be 'true' or 'false'. If it is 'false' the value zero is delivered to direct control to the program segment which checks for conditional actions. If it is 'true', and the list of events is not empty, the effect is to remove an event from the head of the list. Before removing it, appropriate use is made of its 'tim', 'no' and 'data' fields. Simulation time is advanced to the value held as 'tim'. The element is then removed from the stack by the statement:

```
stack:= stack^.link;
```

This causes 'stack' to point to the next element in the list, or to have the value NIL if the element removed was the only one in the list.

The computer program is as follows:

```
PROGRAM simulation(input, output);

TYPE datarr= ARRAY[1..10] OF real;
     pntevent= ^event;
     event= RECORD  tim: real;
                    no: integer;
                    data: datarr;
                    link: pntevent  END;
VAR  time: real;
     get, send: datarr;
     stack: pntevent;
     checked: boolean;
   {Above vars. are part of "package"}
     open: boolean;  patients, i: integer;
     seen: ARRAY[1..4] OF integer;
     docfree: ARRAY[1..4] OF boolean;

PROCEDURE call(i: integer;  t: real);
VAR  occur: real;
     j: integer;
  PROCEDURE putin(VAR e: pntevent);
  VAR  putnew: boolean;
       ee: pntevent;
  BEGIN  putnew:= (e = NIL);
    IF NOT putnew THEN
    putnew:= e^.tim > occur;
```

```
    IF putnew THEN
    BEGIN  ee:= e;  new(e);
      WITH e^ DO
      BEGIN  tim:= occur;  no:= i;
        data:= send;  link:= ee
      END
    END  ELSE  putin(e^.link)
  END  {putin};
BEGIN
  occur:= t + time;
  putin(stack);
  FOR j:= 1 TO 10 DO send[j]:= 0
END {call};

FUNCTION next: integer;
BEGIN IF NOT checked THEN
  BEGIN  next:= 0;  checked:= true
  END
  ELSE  IF stack = NIL THEN next:= 0
  ELSE
  BEGIN  next:= stack^.no;
    get:= stack^.data;
    time:= stack^.tim;
    stack:=stack^.link;
    checked:= false
  END
END {next};

FUNCTION notdone: boolean;
VAR  b: boolean;
BEGIN  b:= (stack <> NIL) OR NOT checked;
  notdone:= b;
  IF NOT b THEN
  BEGIN  writeln(output);
    writeln('End of simulation');
    writeln(output)
  END
END {notdone};

FUNCTION rand: real;  EXTERNAL;

FUNCTION gaussian: real;
VAR  i: integer;  x: real;
BEGIN  x:= 0;
  FOR i:= 1 TO 12 DO
  x:= x + rand;
  gaussian:= (x - 6)
END {gaussian};

FUNCTION negexp: real;
VAR  x: real;
```

```
BEGIN
  x:= rand;  WHILE x = 0 DO x:= rand;
  negexp:= -ln(x)
END;

BEGIN  time:= 0;  stack:= NIL;  checked:= false;
  {These are required as part of package}

open:= true;  patients:= 0;
FOR i:= 1 TO 4 DO
BEGIN  seen[i]:= 0;  docfree[i]:= true;
END;
call(1, 7*negexp);  {arrival of first patient}
call(3, 180);  {shut door in 180 minutes}

WHILE notdone DO
CASE next OF
1: {arrival} IF open THEN
  BEGIN  patients:= patients + 1;
    call(1, 7*negexp)  END;
2: {finish consultation}
  docfree[round(get[1])]:= true;
3: {shut doors}  open:= false;
0: FOR i:= 1 TO 4 DO
  IF docfree[i] AND (patients > 0) THEN
  BEGIN  docfree[i]:= false;
    patients:= patients - 1;
    seen[i]:= seen[i] + 1;
    send[1]:= i;
    call(2, 9 + gaussian)
  END
END {case};

FOR i:= 1 TO 4 DO
writeln('Doctor no.', i:2, ' saw', seen[i]:4,
                ' patients')
END.
```

A run of the program produced the following output:

```
End of simulation

Doctor no. 1 saw  11 patients
Doctor no. 2 saw   7 patients
Doctor no. 3 saw   3 patients
Doctor no. 4 saw   1 patients
```

Once again it must be emphasised that the technique can be applied to situations of much greater complexity.

REFERENCES

1. J.W.J. Williams, "The Elliott Simulation Package", **Computer Journal 6**, pp. 328-331, 1964

2. A.M. Andrew, **Artificial Intelligence** (Abacus, Tunbridge Wells, 1983) pp. 175-177

Chapter 11

OPTIMISATION IN GENERAL

Problems of optimisation can sometimes be solved by the Method of the Calculus. That is to say, derivatives of the objective function with respect to the variables are equated to zero, and the equations solved for the variables. The solutions may correspond to maxima or minima or to points of inflection (or saddle-points if there are two or more variables). Well-known methods using higher derivatives can be used to identify the solutions representing maxima. Solutions subject to constraints of certain kinds can be obtained by the use of undetermined multipliers. (See, for example, Gibson [1]).

Other problems can be solved by the techniques of Linear Programming, as discussed in Chapters 2 to 5. There are also many optimisation problems which cannot be solved by either the Method of the Calculus or by Linear Programming, because the Objective Function and/or the constraints are not linear and not differentiable.

The literature on optimisation methods is enormous; an admirable survey and discussion of them has been written by Schwefel [2]. In this Chapter, two different techniques will be introduced and contrasted. Both of them are of very general applicability, and they operate without the need to differentiate (either numerically or symbolically) either the objective function or functions appearing in the constraints. The first will, for simplicity, be described in a form which does not allow for constraints at all. The second actually needs constraints in order to start.

One example of a problem having a non-differentiable Objective Function arises in connection with the distribution of a commodity from a number of factories to a number of consumers. The commodity passes through a number of depots at which it is transferred from bulk to local transportation. The cost of transportation is proportional to the amount carried and to distance, but at different rates for bulk and local carriage. Given a set of coordinates of customers, as well as their rates of consumption of the commodity, and coordinates of factories and depots, the cost of distribution can be evaluated. There are no capacity restrictions on factories or depots, so each consumer is supplied along the cheapest route.

An optimisation problem arises if some of the factories and depots are considered to be fixed in position, but others are moveable. In a practical application the fixed establishments would be those already built, and the requirement is to decide the most advantageous locations for the others. If there are m moveable entities, the cost of distribution is a function of the 2m variables which are their coordinates. The aim of the optimisation is to choose the 2m values so as to minimise the Objective Function which is the cost.

An application of Rosenbrock's method of optimisation [3] to a problem of this kind is mentioned by Andrew [4]. If the commodity was supplied to each consumer over a fixed route the Objective Function would be everywhere differentiable, but it ceases to be so when a change in a coordinate may cause an alteration in route.

ROSENBROCK'S METHOD

This is an example of a 'hill-climbing' procedure. That is to say, there is at any stage of the optimisation process a set of values of the n variables, and these may be represented by a point in n-dimensional space. The procedure moves this point so as to increase the value of the Objective Function. The hill exists in the (n+1)-dimensional space formed by adding the Objective Function as the extra dimension which can be termed altitude.

Information about the hill is available in the program as a procedure to evaluate the altitude for any set of values of the variables. What might seem the obvious strategy to use would be to determine the altitude at a number of points close to the current operating point and to determine the direction of advance (direction in n-dimensional space) for which the gradient, or rate of increase of altitude, is maximal. One difficulty about this is assigning a suitable meaning to 'closeness' of the points at which the altitude is determined. If they are too close the differences in altitude may be lost, or at least inaccurately represented, because of the finite precision of the digital representation. On the other hand the increments must not be too large since the Objective Function may be strongly non-linear. (By increments is meant the differences in values between the current operating point and the other points at which the altitude is determined.)

It is possible to devise procedures which will automatically scale the increments to suit the situation. Methods which do this are more readily described in terms of the increments, and corresponding changes in Objective Function, rather than in terms of gradients.

A hill-climbing method can depend on making a series of trials at points in the region of the operating point, and simply comparing the Objective Function at each trial point with that at the current point. If the value ul at the new point is higher than that at the current point (u), the current point moves to the new point, otherwise it remains where it was and another trial is made in a different direction. There is still the problem of scaling the increments.

The simplest way to make the trials is by changing one variable at a time. However, the optimisation may come to a stop while not at the optimum if this is strictly adhered to. It is possible to imagine a climber on the top of a ridge running north-east. If he explores in a north-south direction (changing his y-coordinate) he finds he goes downhill. Exploration along his x-coordinate (east-west) is also unrewarding. However, if the ridge slopes upward in the north-east direction, he has a means of increasing his altitude, even though it is not revealed by one-variable-at-a-time exploration.

Rosenbrock finds that, in practical problems, the way to the optimum frequently requires the traversal of steep-sided ridges (or their equivalent in higher-dimensional spaces).

His optimisation method goes through a succession of 'stages', and within each stage it is much like a one-variable-at-a-time method. However, the trial changes are not made along the original coordinate axes (except in the first stage); they are made along a set of orthogonal directions such that one of them is aligned with the total advance of the operating point in the preceding stage. This should mean that one of the directions is approximately aligned with the crest of any ridge to be traversed, and progress should be rapid.

In the program, an array called 'dir' holds a set of n unit vectors indicating the set of orthogonal directions currently in use. Initially, the values in the principal diagonal are set to unity and all other entries to zero, so that the directions correspond to the coordinate axes. At the end of each stage the directions are reassigned by PROCEDURE orth.

For each of the orthogonal directions, a step-size is held in the vector 'step'. These are initially set to the arbitrary value of 0.1. When a step is 'successful', so that the Objective-Function value (ul) at the trial point is greater than or equal to that at the current point, the corresponding step size is increased by a factor alpha, where alpha > 1. If the trial is a 'failure' the step size is multiplied by -beta, where beta < 1. The step-sizes are thus automatically scaled to suit the problem. Purely empirically, Rosenbrock has found that suitable values for alpha and beta are 3 and 0.5, respectively.

In the program, the method is embodied in PROCEDURE rose, which has a function as a parameter. FUNCTION obj is the parameter, and is replaced by FUNCTION eval in the call of PROCEDURE rose. This evaluates the objective function.

A stage is made to end when there has been at least one successful move and one unsuccessful one, in each direction. Two Boolean arrays, called 'succ' and 'fail' are used to determine when this has happened. A new assignment of the orthogonal directions is than made by PROCEDURE orth.

ORTHOGONALISATION

During the operation of a stage, a record is kept in 'prog' of the amount of progress, or shift of the operating point along each of the orthogonal directions. The first thing done by PROCEDURE orth is to fill an array called 'a' with vectors obtained as follows.

The n-th row of 'a' is filled with values representing the advance (as a vector referred to the coordinate axes) along the n-th orthogonal direction. Then the (n-1)-th row of 'a' is filled with the sum of the advances along the n-th and (n-1)-th directions, and so on until the first row is filled with the sum of all the advances and hence with the total progress of the operating point.

The first row of 'a' is copied into the first row of 'dir' and hence becomes the first of the new orthogonal directions. It is scaled to become a unit vector. Then each of the other rows of 'a' is transferred to 'dir', but before it is scaled to become a unit vector it has subtracted from it its projection on to each of the already-established unit vectors. This ensures that the new set of unit vectors is indeed orthogonal. The magnitude of the projection is equal to the scalar product of the two vectors (sum of products of corresponding terms). The vector to be subtracted is the already-established unit vector multiplied by this magnitude.

An alternative method of forming orthogonal vectors has been given by Powell [5]. It does not form the same set of vectors, except that, as required by the overall method, the first one is the same. The advantage of Powell's method is that it lends itself to faster computation. Andrew [4] has reported that, in comparative trials on the goods-distribution problem described earlier, use of Powell's method gave a slightly closer approach to the optimum. It therefore appears to be preferable on grounds other than speed of operation, but it is dangerous to make general conclusions on the strength of a single comparison.

CONSTRAINTS

The method has been described without reference to constraints. Constraints can be taken into account by incorporating them into the Objective Function, in such a way that its value is unaffected as long as no constraint is violated, but falls to a low value whenever there is violation. In principle, the unconstrained method, used with such a composite Objective Function, satisfies the requirement for optimisation subject to constraints.

Unfortunately the modification of the Objective Function as described produces a shape of hill (or response surface) having very steep-sided ridges. In fact, the ridges along the limits imposed by the constraints fall away in vertical cliffs on one side. Rosenbrock's method is better than other hill-climbing methods when it comes to running along steep-sided ridges, but it still fails when faced with this kind of response surface. That is to say, the operating point becomes stationary, or nearly so, while still well away from the optimum.

Rosenbrock realised the difficulty, and modified the way of forming the composite Objective Function so that the ridges at the edges of the permitted area were not razor-sharp. He also arranged that the low value of the Objective Function outside the permitted region was altered as optimisation proceeded, and that the cross-section of the tip of the ridge in boundary areas was altered, by a highly ingenious expedient, during the run. Other workers have made other modifications of the method, as reviewed by Box [6].

Despite all these efforts, however, Rosenbrock's method still leaves something to be desired when applied to problems involving constraints. It is acknowledged by Box to be better than the others with which he compares it in unconstrained situations. Of course, conclusions of this kind must be somewhat imprecise since, as has been seen, the same problem may be presented either as one involving constraints or as one with a more complex Objective Function and no constraints. However, according to customary methods of presentation of problems, Rosenbrock's method seems to perform well on unconstrained ones.

AN APPLICATION

To illustrate the method, optimisation has been used to solve the pair of simultaneous equations:

$$x_1 + 2x_2 = 7$$

$$2x_1 + x_2 = 5$$

Since the program has been written so as to maximise the Objective Function, it is calculated as the negative of the sum of squares of the imbalances of the equations:

$$u = -(x_1 + 2x_2 - 7)^2 - (2x_1 + x_2 - 5)^2$$

The variables have to be set to initial values, and these are, arbitrarily, $x_1 = x_2 = 0$. The solution is $x_1 = 1$, $x_2 = 3$.

STOPPING CRITERION

In the run of the program the number of stages was arbitrarily set to six. This allows comparison with a result quoted by Rosenbrock, and for some reason the convergence on the optimum was not as good as Rosenbrock found.

The difference in performance was presumably due to minor differences in implementation, perhaps in the precision of number-representation in the two computers used. The present implementation required two further stages (making a total of eight) to give a result as close as Rosenbrock quotes for six. The overall method is found to be 'robust' in that minor changes do not prevent its eventual convergence. However, its speed of convergence is readily affected by small changes, in ways which are very difficult to explain.

Ideally, an optimisation method should embody a stopping criterion so that it does not waste time when no further improvement is possible. Rosenbrock suggests that the program should evaluate, at the end of each stage, the amount of progress made during it, as the square root of the sum of squares of the first row of the array 'a' computed within PROCEDURE orth. In Rosenbrock's notation this is denoted by $|A_1|$. He also suggests evaluation of the same function of the second row, denoted by $|A_2|$, being the amount of progress in directions other than the main one. The significant criterion is then the ratio of these two amounts of progress:

$$|A_2|/|A_1|$$

If this is less than 0.3 it is likely that further improvement is possible. The program prints $|A_1|$ and the ratio.

COMPUTER PROGRAM

The computer program is as follows:

```
PROGRAM rosenbrock(input, output);

TYPE vec= ARRAY[1..10] OF real;
     mat= ARRAY[1..10,1..10] OF real;

VAR  alpha, beta: real;

PROCEDURE orth(prog: vec;  VAR dir: mat;
                     VAR b1, b2: real;  n: integer);
VAR  a: mat;
     i, ii, j: integer;
     x: real;
BEGIN  FOR i:= 1 TO n DO  FOR j:= 1 TO n DO
                                        a[i, j]:= 0;
  FOR j:= 1 TO n DO a[n, j]:= prog[n]*dir[n, j];
  FOR i:= n-1 DOWNTO 1 DO  FOR j:= 1 TO n DO
    a[i, j]:= a[i+1, j] + prog[i]*dir[i, j];
  FOR i:= 1 TO n DO
  BEGIN  FOR j:= 1 TO n DO dir[i, j]:= a[i, j];
    FOR ii:= 1 TO i-1 DO
    BEGIN  x:= 0;
      FOR j:= 1 TO n DO x:= x+a[i, j]*dir[ii, j];
      FOR j:= 1 TO n DO
            dir[i, j]:= dir[i, j]-x*dir[ii, j]
    END;
    x:= 0;
    FOR j:= 1 TO n DO x:= x+sqr(dir[i, j]);
    x:= sqrt(x);
```

```
     {In PASCAL, sqr means square, sqrt means
                                    square root}
    FOR j:= 1 TO n DO dir[i, j]:= dir[i, j]/x
  END;
  x:= 0;  FOR j:= 1 TO n DO x:= x+sqr(a[1, j]);
  bl:= sqrt(x);
  x:= 0;  FOR j:= 1 TO n DO x:= x+sqr(a[2, j]);
  b2:= sqrt(x)
END {orth};

FUNCTION eval(x: vec): real;
BEGIN
  eval:= -sqr(x[1]+2*x[2]-7)-sqr(2*x[1]+x[2]-5)
END;

PROCEDURE rose(n, nstages: integer;
                         FUNCTION obj(x: vec): real);
LABEL 1;
VAR  x, xl, step, prog: vec;
     dir: mat;
     succ, fail: ARRAY[1..10] OF boolean;
     i, j, stage, trial: integer;
     u, ul, bl, b2: real;
     endstage: boolean;
BEGIN  trial:= 0;
  FOR i:= 1 TO n DO x[i]:= 0;
  u:= obj(x);
  FOR i:= 1 TO n DO step[i]:= 0.1;
  FOR i:= 1 TO n DO FOR j:= 1 TO n DO
    IF i=j THEN dir[i, j]:= 1
    ELSE  dir[i, j]:= 0;
  FOR stage:= 1 TO nstages DO
  BEGIN  FOR i:= 1 TO n DO
    BEGIN  succ[i]:= false;  fail[i]:= false;
      prog[i]:= 0  END;
    endstage:= false;
    WHILE NOT endstage DO
    BEGIN  FOR i:= 1 TO n DO
      BEGIN  trial:= trial+1;
        FOR j:= 1 TO n DO
                    xl[j]:= x[j]+step[i]*dir[i, j];
        ul:= obj(xl);
        IF ul>=u THEN
        BEGIN  succ[i]:= true;  u:= ul;
          FOR j:= 1 TO n DO x[j]:= xl[j];
          prog[i]:= prog[i]+step[i];
          step[i]:= alpha*step[i]
        END ELSE
        BEGIN  fail[i]:= true;
          step[i]:= -beta*step[i]
        END;
```

```
        endstage:= true;
        FOR j:= 1 TO n DO
        IF NOT (succ[j] AND fail[j]) THEN
                                          endstage:= false;
        IF endstage THEN GOTO 1
      END;
    1: END;
    orth(prog, dir, b1, b2, n);
    writeln('Stage', stage:3, '  Trials', trial:5);
    writeln('Progress in stage', b1:9);
    writeln('Ratio 1st 2 compts', b2/b1:9);
    writeln('Objective fn.', u:9);
    writeln('Variables:');
    FOR j:= 1 TO n DO writeln(x[j]:9);
    writeln(output)
  END {of stage}
END {rose};

BEGIN  alpha:= 3;  beta:= 0.5;
  rose(2, 6, eval)
END.
```

The output was as follows:

```
Stage  1  Trials   10
Progress in stage 4.00E+00
Ratio 1st 2 compts 1.00E+00
Objective fn.-2.11E+00
Variables:
-5.00E-02
 4.00E+00

Stage  2  Trials   22
Progress in stage 2.83E-01
Ratio 1st 2 compts 4.47E-01
Objective fn.-1.53E+00
Variables:
 7.97E-02
 3.75E+00

Stage  3  Trials   27
Progress in stage 1.57E+00
Ratio 1st 2 compts 2.43E-01
Objective fn.-5.90E-01
Variables:
 1.11E+00
 2.57E+00
```

```
Stage  4  Trials   32
Progress in stage 5.13E-01
Ratio 1st 2 compts 5.55E-01
Objective fn.-7.29E-02
Variables:
 1.05E+00
 3.08E+00

Stage  5  Trials   45
Progress in stage 1.09E-01
Ratio 1st 2 compts 9.83E-01
Objective fn.-6.07E-03
Variables:
 9.42E-01
 3.05E+00

Stage  6  Trials   51
Progress in stage 2.50E-02
Ratio 1st 2 compts 8.00E-01
Objective fn.-2.58E-03
Variables:
 9.63E-01
 3.03E+00
```

BOX'S METHOD

Rosenbrock's method was produced after a great deal of experimentation, and comparison with earlier methods. Nevertheless, both it and its subsequent developments do not perform well on some problems with constraints. Box [6] has reviewed the situation and produced a method which usually performs better when the problem involves constraints. He keeps the constraints separate from the Objective Function.

Box's method does not correspond to the ascent of a hill by a single climber. Instead, he has a set of current points distributed over the permitted region of the n-dimensional space, like a group of climbers having some means of intercommunication.

The method is developed from an earlier one using a 'simplex' of points. (The word simplex is used here in a mathematical sense having no connection with the Simplex Method of Linear Programming.) A simplex, in n-dimensional space, is a set of points, (n+1) in number, and equidistant from each other. In 2-dimensional space the points are vertices of an equilateral triangle, in 3 dimensions they are corners of a regular tetrahedron, and so on. An optimisation method which has been used involves evaluating the Objective Function at each of the points of a simplex, and then modifying the simplex by eliminating the point which is 'worst' in the sense of having the poorest evaluation. A new point is added which is the reflection of the eliminated one in the centroid of the remaining points, so that a simplex is again formed. Repetition of this procedure causes the simplex to move towards the optimum.

This method has some drawbacks, the main one being that the simplex is fixed in size. For best performance it should be scaled to suit the problem situation, with a means of adjusting the size as the optimum is approached.

Box uses an irregular cluster of points which he terms a 'complex'. He denotes the number of points by k, and he finds it best to let k be considerably greater than (n+1) as required for a simplex, where n is, as before, the dimensionality of the optimisation. Empirically, Box has found that a good value for k is 2n.

He divides constraints into two categories. The first is the category of 'explicit' constraints, which simply consist of a pair of fixed limits, upper and lower, for each of the variables. All constraints not in this simple form are 'implicit', and for a specific problem a section of program is needed (FUNCTION legal, in the program shown here) to indicate whether or not a given point is permitted by these constraints.

The optimisation proceeds by finding the worst point in the complex and replacing it by its reflection in the centroid of the others. To make the complex grow larger, the reflection is made to be an over-reflection. That is to say, the distance of the new point from the centroid is greater than the distance of the old point by a factor of alpha, where alpha > 1. Box has found 1.3 to be a suitable value for alpha.

It will sometimes happen that the new point contravenes one or more constraints. If it contravenes an explicit constraint, the value of the variable to which the constraint refers is replaced by a value 0.000001 inside the constraint value. This is done for all n variables. If the point then contravenes an implicit constraint it is replaced by a point half-way between it and the centroid of the remaining points, and so on until an acceptable point is found. (In the program, this drawing-back towards the centroid is allowed to happen 20 times, and if the point is still impermissible it is replaced by a copy of the centroid. It has been found that the effect of rounding-off errors is such that the drawing-back process does not always converge exactly on the centroid, and the point may remain impermissible even though the centroid is in the permitted region.)

If the point found to be the worst is the one most recently introduced, it is again replaced by a point half-way between it and the centroid, until a different point becomes worst.

The drawing-back of points towards the centroid causes the complex to shrink in size, but the use of over-reflection expands it. Once it comes to be clustered around the optimum the shrinking effect predominates, and the complex shrinks on to the optimum. The process is terminated when five successive evaluations of the Objective Function give results which are indistinguishable within the finite resolution of the digital representation. In the program the number of evaluations giving indistinguishable results is counted in the integer variable 'count'.

The method might fail to find the optimum if, at some stage, all the points came to lie exactly in a subspace of lower dimensionality. For example, if the optimisation was in three dimensions, it could happen that all the points came to lie in a plane, and subsequent over-reflection in a centroid, or drawing-back towards a centroid, would not produce points outside the plane. It is found in practice that such collapse into a subspace does not happen. The use of over-reflection means that any slight deviation from the subspace is amplified, and slight deviations can be introduced by rounding-off errors or by the adjustment of points to fall within explicit constraints.

The set of k points forming the complex can be chosen automatically with the help of the pseudo-random number generator provided in the system. It is necessary to let the first point be assigned manually so as to conform to all the constraints.

Then for each coordinate of each subsequent point, a value is assigned as:

l + (h - l)*rand

where 'rand' is a call of the p.r.n.g. delivering a value in the range 0 to 1, and h and l are the upper and lower limits constituting the explicit constraints on this variable.

If the point chosen in this way is found to contravene an implicit constraint it is replaced by a point half-way between it and the centroid of the points already chosen, until a permissible point is found.

COMPUTER PROGRAM

A program implementing Box's method is as follows. It has been set to solve the same problem as the Rosenbrock method. Both of the explicit constraints have been set to limit the variables to values between -10 and +10, and the first point of the complex is set initially to (0, 0). The FUNCTION legal always delivers 'true' since there are no implicit constraints.

```
PROGRAM box(input, output);

TYPE   vec= ARRAY[1..10] OF real;
       point= RECORD  x: vec;
                      u: real  END;

VAR    limits: ARRAY[1..10,1..2] OF real;
       n, k, tryleg, trial, count,
                           i, j, m, ml: integer;
       complex: ARRAY[1..20] OF point;
       centroid: vec;
       last, alpha: real;

FUNCTION eval(x: vec): real;
BEGIN
  eval:= -sqr(x[1]+2*x[2]-7)-sqr(2*x[1]+x[2]-5);
END {eval};

FUNCTION legal(x: vec): boolean;
BEGIN  legal:= true  END {legal};

FUNCTION check(p: point): boolean;
```

```
BEGIN  check:= legal(p.x);
  tryleg:= tryleg+1
END  {check};

PROCEDURE evaluate(VAR p: point);
VAR v: real;
BEGIN  v:= eval(p.x);  p.u:= v;
  trial:= trial+1;
  IF v=last THEN count:= count+1
    ELSE count:= 0;
  last:= v
END {evaluate};

FUNCTION rand: real;  EXTERNAL;

PROCEDURE fillvec(VAR pt: vec);
VAR i: integer;
    l, h: real;
BEGIN  FOR i:= 1 TO n DO
  BEGIN  l:= limits[i, 1];
    h:= limits[i, 2];
    pt[i]:= l+(h-l)*rand
  END
END  {fillvec};

PROCEDURE drawin(VAR x: vec; y: vec);
VAR i: integer;
BEGIN  FOR i:= 1 TO n DO
  x[i]:= (x[i]+y[i])/2
END  {drawin};

PROCEDURE makecen(no, exclude: integer);
{Forms centroid of first no points,
                                excluding one}
VAR i, j: integer;
BEGIN  FOR j:= 1 TO n DO centroid[j]:= 0;
  FOR i:= 1 TO no DO
  IF i <> exclude THEN
    WITH complex[i] DO
    FOR j:= 1 TO n DO
    centroid[j]:= centroid[j]+x[j];
  FOR j:= 1 TO n DO
    centroid[j]:= centroid[j]/(no-1)
END  {makecen};

FUNCTION findmin: integer;
VAR  min: real;
     i, m: integer;
BEGIN  m:= 1;  min:= complex[1].u;
  FOR i:= 2 TO k DO
  IF complex[i].u < min THEN
```

```
  BEGIN  min:= complex[i].u;  m:=i
  END;
  findmin:= m
END {findmin};

BEGIN  n:= 2;  alpha:= 1.3;
  tryleg:= 0;  trial:= 0;  last:= 0;
  limits[1, 1]:= -10;  limits[1, 2]:= 10;
  limits[2, 1]:= -10;  limits[2, 2]:= 10;
  WITH  complex[1] DO
  BEGIN  x[1]:= 0;  x[2]:= 0  END;
  k:= 2*n;

  {Now to form initial complex of points}
  evaluate(complex[1]);
  FOR i:= 2 TO k DO
  BEGIN  fillvec(complex[i].x);
    WHILE NOT check(complex[i]) DO
    BEGIN  makecen(i, i);
      drawin(complex[i].x, centroid)
    END;
    evaluate(complex[i])
  END;

m:= findmin;  ml:= m;

REPEAT  makecen(k, m);
  WITH complex[m] DO
  FOR j:= 1 TO n DO
  BEGIN
    x[j]:= centroid[j]-alpha*(x[j]-centroid[j]);
    IF x[j] < limits[j, 1] THEN
                   x[j]:= limits[j, 1] + 1E-6
    ELSE IF x[j] > limits[j, 2] THEN
                   x[j]:= limits[j, 2] - 1E-6
  END;
  i:= 0;
  WHILE NOT check(complex[m])  DO
  BEGIN  i:= i+1;
    IF i < 20 THEN
    drawin(complex[m].x, centroid)
    ELSE  complex[m].x:= centroid
  END;
  evaluate(complex[m]);
  REPEAT  m:= findmin;
    IF m=ml THEN
    BEGIN  drawin(complex[m].x, centroid);
      evaluate(complex[m])
    END;
  UNTIL  (m <> ml) OR (count >= 5)
                          OR (trial > 1000);
```

```
  m1:= m;
UNTIL  (count >= 5) OR (trial > 1000);

writeln('No.  of evaluations', trial:4);
writeln('No.  of checks', tryleg:4);
writeln('Objective fn.', last);
writeln('Variables:');
WITH complex[1] DO
FOR j:= 1 TO n DO writeln(x[j])
END.
```

Running this program gave the following output:

```
No. of evaluations 234
No. of checks 116
Objective fn.-2.0236257E-11
Variables:
 1.0000004E+00
 2.9999977E+00
```

To solve the Linear Programming example used to introduce the Simplex Method in Chapter 2, FUNCTION eval and FUNCTION legal were rewritten as follows:

```
FUNCTION eval(x: vec): real;
BEGIN
  eval:= 3*x[1]+2*x[2]
END {eval};

FUNCTION legal(x: vec): boolean;
BEGIN  legal:= (x[1]+x[2] <= 10) AND
  (2*x[1]+x[2] <= 16) AND (x[1] + 2*x[2] <= 16)
END {legal};
```

The limits for the explicit constraints were set to be from 0 to 10 instead of from -10 to +10, and the first point was initialised to (0.1, 0.1). The output was as follows:

```
No. of evaluations 100
No. of checks 183
Objective fn. 2.5999972E+01
Variables:
 6.0000198E+00
 3.9999574E+00
```

This is in good agreement with the result obtained by Linear Programming, where the optimum profit was 26 units and values of the variables were 6 and 4 respectively.

CONCLUSION

The methods described in this Chapter are perhaps better known in connection with Engineering than for their applications in Operations Research. They have applications in engineering design, and are also interesting as models for the continuous automatic optimisation underlying self-optimising control systems. They may also be useful models of biological learning and therefore have a bearing on self-organisation in living systems including the nervous system. To pursue these ideas further would mean going well outside the scope of the present book, but into a fascinating area which will be treated in a later one.

Statisticians sometimes need to solve problems of optimisation and when they do they usually refer to the early work of Robbins and Monro - see Schwefel [2] p. 18. In at least some of their applications, more recent developments including the two treated here may be useful.

The methods are of very general applicability, particularly since they can be applied where the Objective Function is evaluated in ways other than by evaluation of a straightforward mathematical function. The evaluation could, for example, depend on a run of a computer simulation, or on experiments on an actual physical process. Optimisation methods to be used with actual process plants have been studied for a long time under the heading of Industrial Experimentation - see for example Davies [7].

In the Operations Research context these methods provide something of a 'block-buster' approach which can sometimes produce results where all else has failed. As such, they seem to be a fitting topic on which to end this book. I hope it has proved interesting as well as useful.

REFERENCES

1. G.A. Gibson, **Elementary Treatise on the Calculus** (Macmillan, London, 2nd Ed. 1940), pp. 412-416

2. H.-P. Schwefel, **Numerical Optimization of Computer Models** (Wiley, Chichester, 1981)

3. H.H. Rosenbrock, "An automatic method for finding the greatest or least value of a function." **Computer Journal 3**, pp. 175-184, 1960

4. A.M. Andrew, "The calculation of orthogonal vectors" (Letter to the Editor). **Computer Journal 12**, p. 411, 1969

5. M.J.D. Powell, "On the calculation of orthogonal vectors." **Computer Journal 11**, pp. 302-304, 1968

6. M.J. Box, "A new method of constrained optimization and a comparison with other methods." **Computer Journal 8**, pp. 42-52, 1965

7. O.L. Davies (ed.) **Design and Analysis of Industrial Experiments** (Oliver and Boyd, Edinburgh, 1956)

AUTHOR INDEX

Unbracketed numbers refer to the chapters; numbers in brackets give the number of the reference in the chapter.

FURTHER READING

Some of the books suggested here are also included in the References, but others have not been previously mentioned.

CHAPTER 1

The following give background information on Operations Research:

R.L. Ackoff, **The Art of Problem Solving, Accompanied by Ackoff's Fables** (Wiley, New York, 1978)

S. Beer, **Brain of the Firm** (Professional Press, London, 1972)

A great many popular books on BASIC have appeared in recent years, most of them intended for use with particular small computers. Also, manuals are supplied with the computers, that included with the Sinclair SPECTRUM being:

S. Vickers, **ZX SPECTRUM BASIC Programming** (Sinclair, Cambridge, 2nd ed. 1983)

The standard work on the language, with many interesting sample programs, is:

J.G. Kemeny and T.E. Kurtz, **Basic Programming** (Wiley, New York, 3rd ed. 1980)

The PASCAL language is also treated in a number of proprietary works. It owes its origin to Wirth, and is the subject of a British Standard. It is treated comprehensively in each of the following:

L. Atkinson, **Pascal Programming** (Wiley, Chichester 1980)

K. Jensen and N. Wirth, **Pascal - User Manual and Report** (Springer, New York 1978)

I.R. Wilson and A.M. Addyman, **A Practical Introduction to Pascal - with BS6192** (Macmillan, London, 2nd ed. 1982)

CHAPTERS 2-5

There is an extensive literature on Linear Programming, including many good textbooks. The following can be recommended, but so also could a great many others:

W.W. Garvin, **Introduction to Linear Programming** (McGraw-Hill, New York, 1960)

B. Krekō, **Linear Programming** (Pitman, London, 1968)

T.H. Naylor, E.T. Byrne and J.M. Vernon **Introduction to Linear Programming: Methods and Cases** (Wadsworth, Belmont, California, 1971)

S. Vajda, **An Introduction to Linear Programming and the Theory of Games** (Methuen, London 1960 and 1963, and in 1966, 1970 and 1978 in Methuen Science Paperback Series)

Linear Programming is also treated in a number of books which, like this one, cover a range of O.R. topics. Among these may be mentioned:

R.T. Eddison, K. Pennycuick and B.H.P. Rivett (eds.) **Operational Research in Management** (E.U.P., London, 1962)

F.S. Hillier and G.J. Lieberman **Operations Research** (Holden-Day, San Francisco, 2nd ed. 1974)

L.J. Krajewski and H.E. Thompson, **Management Science: Quantitative Methods in Context** (Wiley, New York, 1981)

M. Sasieni, A. Yaspan and L. Friedman **Operations Research: Methods and Problems** (Wiley, New York, 1959)

R.J. Thierauf, **An Introductory Approach to Operations Research** (Wiley, Santa Barbara, 1978)

F.M. Wilkes, **Elements of Operational Research** (McGraw-Hill, Maidenhead, 1980)

CHAPTER 6

Since Critical Path Analysis is a matter of operations on graphs, the reader may wish to know more about Graph Theory. The following books introduce this topic, but quickly go well beyond anything needed to understand the treatment of C.P.A. presented here:

C. Berge, **The Theory of Graphs and its Applications** (Methuen, London, 1962)

A. Kaufmann, **Points and Arrows** (Transworld, London, 1972)

Critical Path Analysis is discussed, with a program in BASIC implementing it, in:

J.G. Kemeny and T.E. Kurtz **Basic Programming** (Wiley, New York, 3rd ed. 1980) pp. 183-187, or 2nd ed., 1971, pp. 94-96

It is also discussed in the general works on O.R. by Hillier and Lieberman, Krajewski and Thompson, Thierauf, and Wilkes, listed above as applicable to Chapters 2-5.

CHAPTER 7

Bellman's original book may be found to be rather indigestible, and applicable to areas rather remote from those considered here. A later one discussing interesting applications is:

R.E. Bellman and S.E. Dreyfus, **Applied Dynamic Programming** (Princeton Univ. Press, Princeton, 1962)

Dynamic Programming is also treated in all six of the general works on O.R. listed as applicable to Chapters 2-5.

CHAPTER 8

Convenient techniques for complicated continuous simulations have been developed by Forrester, whose book should be consulted for details:

J.W. Forrester, **Industrial Dynamics** (M.I.T. Press, Cambridge, Mass. 1961)

CHAPTERS 9 and 10

Techniques for discrete-event simulation of stochastic situations are discussed in all six of the general works on O.R., listed as applicable to Chapters 2-5.

CHAPTER 11

A comprehensive review of optimisation methods is given in:

H.-P. Schwefel, **Numerical Optimization of Computer Models** (Wiley, Chichester, 1981)

It is also treated in Chapter 9 of Krajewski and Thompson, listed above as applicable to Chapter 2-5.

SUBJECT INDEX

(Numbers refer to pages)